AN INTRODUCTION TO
COMPUTER-AIDED DESIGN AND
MANUFACTURE

COMPUTER SCIENCE TEXTS

COMPUTER SCIENCE TEXTS

An Introduction to Computer-Aided Design and Manufacture

MARTIN J. HAIGH
CEng, MIMech.E, MIMI, MBCS
Lecturer in Computing,
Department of Mathematics and Computing
The Percival Whitley College of
Further Education, Halifax

BLACKWELL SCIENTIFIC PUBLICATIONS

OXFORD   LONDON   EDINBURGH

BOSTON   PALO ALTO   MELBOURNE

© 1985 by
Blackwell Scientific Publications
Editorial offices:
Osney Mead, Oxford, OX2 0EWL
8 John Street, London, WC1N 2ES
9 Forrest Road, Edinburgh, EH1 2QH
52 Beacon Street, Boston
  Massachusetts 02108, USA
667 Lytton Avenue, Palo Alto
  California 84301, USA
107 Barry Street, Carlton,
  Victoria 3053, Australia

First published 1985

Phototypeset by
Oxford Computer Typesetting

Printed and bound in
Great Britain by
Redwood Burn Limited,
Trowbridge, Wiltshire

British Library Cataloguing in
Publication Data

Haigh, Martin
  An introduction to computer-aided
  design and manufacture.—(Computer
  science texts)
  1. Engineering design—Data processing
  I. Title. II. Series
  620'.00425'02854   TA174

  ISBN 0–632–01242–0

Distributed in North America by
Computer Science Press, Inc.,
11 Taft Court,
Box 6030, Rockville,
Maryland 20850, USA

Contents

Preface

Computer-aided design and manufacture is probably the most significant advance in engineering design and manufacture in modern times. CAD/CAM will have a major effect on manufactured products by the 1990s and will influence the economics of the manufacturing nations in the next century. Surveys estimating the growth of CAD/CAM state that by 1990 CAD techniques will be used for at least half of the new assemblies designed, and that in the USA the number of CAD/CAM installations grows by more than 30% per year.

A report to the Engineering board of the Science and Engineering Research Council by an engineering design working party in June 1983 includes the following points in its recommendations:

● design is the very core of engineering and the education of engineers should reflect this;

● modern graduates should be trained in the traditional design skills and the use of modern electronic aids such as CAD/CAM;

● the appropriate bodies should be urged to validate only those engineering courses which have this essential design orientation;

● CAD/CAM will almost certainly improve the status of designers and will change their role in industry.

It seems clear that the use of CAD/CAM is expanding in industry in the USA, Japan, West Germany and, as far as can be judged, in the UK also, and this expansion will continue, at least to the end of the century. It is also evident that CAD/CAM is becoming an essential part of modern engineering and computing courses.

The purpose of this book is to provide students and executives in industry with an appreciation of what a computer-aided design or manufacturing system is, and how it works. The hardware and software elements of a CAD/CAM system are described and the uses of these elements are discussed. Case studies give the reader a step-by-step account of how typical CAD/CAM systems operate. Individual chapters describe the application of computers in design, draughting and three-dimensional modelling. The use of the finite element method as a design/analysis tool is outlined. The use of computers in manufacturing is discussed, followed by two separate chapters on machine tool control and robotics. The benefits to be derived from computer-aided design and manufacturing systems are outlined and a few predictions for the future of CAD/CAM are given. Finally an appendix is included listing some of the major CAD/CAM system suppliers to give an indication of systems currently available.

I would like to thank the many companies who have contributed material or who have given permission to mention their systems in this book. In particular, Mr A.M. Denford of Denford machine tools, Mr D. Oxley of Micro Aided Engineering, Dr R.D. Henshell of Pafec Limited, and Dr M. Susan Bloor of the Geometric Modelling Project at the University of Leeds, for allowing me to quote their systems in the case studies. My sincere thanks go to Mr N.K. Shaw and Dr N.C. Wallbridge of the University of Leeds for their valuable suggestions on selected chapters. I also wish to thank Prof B. Parsons of Queen Mary College, University of London for his comments and for his help during the early part of my research work at Leeds University. I am indebted to two people for their help during the production of this book: to Mr Mark Wisniewski for drawing the line diagrams in chapters 6, 8 and 9, and to my wife, Melanie, for her continued support throughout the writing of this book and for typing the entire manuscript.

M.J.H.

Introduction

With the increased use of computers, the engineer has acquired a very capable assistant. Not only can the computer perform calculations rapidly, it can do many more tasks which make it an invaluable aid, and consequently is changing the way of life for the engineer.

The Computer can also:

(i) Carry out complicated mathematical operations without error.
(ii) Store vast amounts of data on magnetic media.
(iii) Retrieve this data very quickly.
(iv) Produce detailed drawings to very great accuracy.
(v) Print good quality text at high speed.
(vi) Converse with the Engineer in many. ways, thereby providing a man–machine interaction.

The use of computers in Engineering Design and Manufacture brings with it many benefits as there are a number of tasks for which the computer is more able and better equipped to perform than man.

The Department of Trade and Industry is seeking to promote the acceptance and application of Computer-Aided Design and Manufacture (CAD/CAM) and is striving to improve industry's awareness of this exciting technology. There is much confusion as to what exactly CAD/CAM is, and who is to benefit from it.

CAD and CAM are only a part of a much broader technology, *Computer-Aided Engineering* (CAE).

CAE encompasses other important requirements such as Engineering analysis and plant control. CAE is a very powerful tool and it offers industry the ability to save time and tedium in the Drawing Office, produce better designs and faster quotations and to provide automatic machine tool control from a design produced on a computer. With the use of specially prepared pre-written software (packages) to run on the computer, this offers an aid to engineers of all disciplines, and when integrated with other systems CAE becomes a comprehensive tool to form a link throughout an engineering company. Figure 1 shows the main activities of Computer-Aided Engineering.

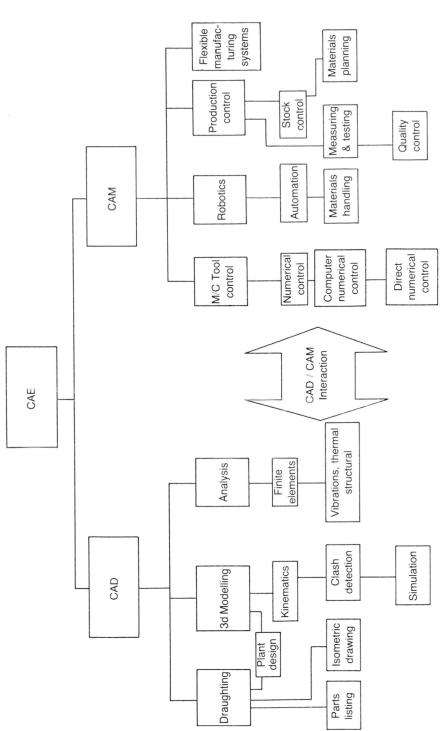

Fig. 1. The main activities of Computer-Aided Engineering

Acknowledgements

The following organizations have kindly given permission to reproduce
   material in this book:

A M Admel
Babcock Fata Limited
Benson Electronics Limited
Binks-Bullows Limited
British Robot Association
CAD CAM International
Cadsource Limited
Cascade Graphics Development
Cambridge Interactive Systems
   Ltd
Comtec Data Systems Limited
Computervision
Dainichi-Sykes Robotics Ltd
Delta C.A.E. Limited
Denford Machine Tools Ltd
Fairey Automation Ltd
Further Education Unit
G.E.C. Electrical Projects Ltd
Hepworth Engineering Ltd
Intergraph (Great Britain) Ltd
University of Leeds — Geometric
   Modelling Project

Micro Aided Engineering Ltd
Pafec Limited
Precom
Production Engineering Research
   Association
Radan Computational
Robocom Ltd
Saab-Scania, Sweden
Shape Data Ltd
S.I.A. Computer Services
Siemens Control and Automation
   Systems
Sulzer Bros., Switzerland
Tektronix U.K. Limited
Traub Ltd
Unimation (Europe) Ltd
Veritec Limited
Versatec Electronics Limited
V.G. Systems Limited
V.S. Technology Group
Watanabe
Westland Helicopters Limited

A few of the entries in Appendix 1 are extracts from the November
1984 edition of *CAD CAM International* published by EMAP Business
and Computer Publications Ltd.

Chapter 1

What is Computer-Aided Design?

1.1 INTRODUCTION

In the broadest sense, Computer-Aided Design refers to any application of a computer to the solution of design problems. More specifically, CAD is a technique in which the engineer and a computer work together as a team, utilizing the best characteristics of each.

The engineer may communicate with the computer in many forms, either via the visual display screen, keyboard, graph plotter or many more man–machine interfaces. The engineer can ask a question and receive an answer from the computer in a matter of seconds.

This chapter takes a brief look at the design process and discusses how the computer can be used to aid design. The various applications of computers to the engineering design process are then outlined.

1.2 THE DESIGN PROCESS

Computer-aided design enables the engineer to test a design idea and to rapidly see its effect, the design idea can then be modified and re-assessed. The process being repeated until a good design is achieved. Figure 1.1 illustrates this iterative design process. Following each iteration, the design solution hopefully improves. Therefore, the more cycles that can be carried out within the constraints of time and money, the better the result should be. The computer enables the engineer to examine the results of his design analysis very rapidly so he can amend the design very easily and quickly to achieve a design solution within the financial, material and time constraints.

1.3 HOW THE COMPUTER AIDS DESIGN

In the past, the conventional tools of the engineer in his/her role as a designer, have been drawing boards and instruments, calculators and technical data sheets. More recently, however, these conventional tools are being replaced by *digital* computer equipment. The engineer now has access to substantial computing power and these computer resources

1

Chapter 1

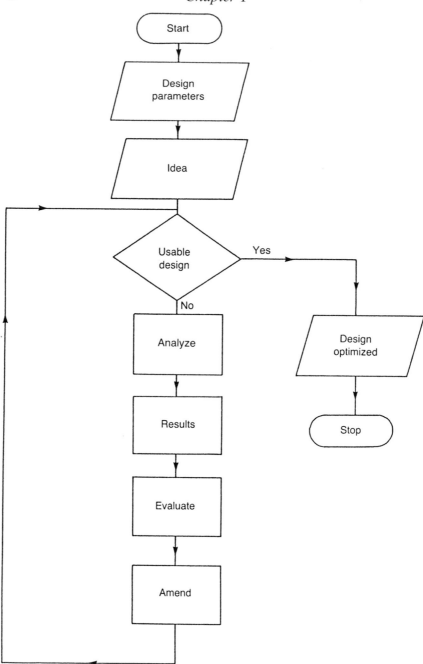

Fig. 1.1. The iterative design process

can be exploited to speed up and improve the accuracy of the design process.

The computer will perform large numbers of complicated calculations in a very short space of time and will produce results which are accurate and reliable. This foregoing feature of the computer proves invaluable in its role as a design aid as the number of calculations required by some designs could simply not be performed by man in a reasonable time.

The computer is capable of holding vast quantities of information on permanent media such as magnetic disc or temporarily in immediate access store.

It is therefore possible to represent the details of an engineering drawing or the shape of a car body in digital form and store this digital information in memory. This data can then be retrieved from memory, rapidly converted and displayed on a VDU graphics screen, or alternatively, plotted onto paper using a graph plotter.

Engineers can therefore have *immediate recall* of say a set of engineering drawings for a given machine so that they can quickly and easily update or amend any part of the drawing. The drawing data can then be written back to memory in its updated form.

1.4 THE USES OF CAD

Computer-Aided Design has many roles to play in the world of engineering, such as: the application of computer systems to the production of engineering drawings; the use of finite element techniques to solve stress and thermal problems on complex components; the analysis of mechanisms and linkages, plus a host of additional engineering applications.

The Computer in the Drawing Office

The standard method of communication between engineers has, for many years, been via engineering drawings. Engineers have traditionally developed their designs by firstly sketching out an idea, re-drawing the design in an improved form then going through the time-consuming, laborious process of producing a working drawing, and finally producing a parts list. The computer can now play a significant role in this development process by working from a limited design specification to produce a drawing which can be examined and amended as necessary. The designer inputs the design parameters (overall co-ordinates, holes, grooves, etc.), together with any special requirements, then the computer uses standard pre-stored information to produce the initial drawing for assessment.

Typical capabilities of a draughting package might include:

(i) Production of orthographic three-view engineering drawings on standard drawing sheets. These drawings can be fully dimensioned and toleranced and can include hatched sections as well as hidden line detail.

(ii) Production of design layouts and general assemblies.

(iii) Generation of materials and parts lists.

(iv) Production of precision outlines, such as those required for shadow projection inspection and optical profile grinding.

(v) Production of diagrams and symbols of all kinds including wiring, hydraulics, electronics, pneumatics, printed circuit layouts and artwork masters.

Many computer-aided draughting systems are capable of producing isometric views and exploded assemblies which might be required for manuals or handbooks. Some systems also offer process and plant layout.

Once a drawing has been created and stored in the system, it is possible to specify the scale the drawing is to be plotted to. This can be a useful facility when drawings of reduced size are required for service and maintenance manuals. It is also useful when precision outlines are required for inspection purposes.

The Computer as a Modelling Aid

There are some design operations where 2-dimensional drawings are simply not adequate to convey the desired design information and expensive 3-dimensional clay or wood models have to be created. The engineer can eliminate the costly task of building prototypes by representing the design as a software model using a *three-dimensional modelling system*. There are various types of 3-D modellers ranging from the simple wireframe type, which represents the object as a collection of lines to give a pictorial view of the component, to the more sophisticated solid modellers where the object is represented as a 3-D solid and from which complete information about the component can be extracted. For example, volume of material required, centroids, centre of gravity, moments of inertia and so on.

3-D models can usually be rotated on the screen, shaded, coloured, or viewed from any angle. It is the extensive visualization features of modellers that render them ideal tools for the marketing department. 3-D modellers are relatively expensive and whilst some engineering companies may have difficulty in justifying such a luxury, there are situations where these systems prove invaluable.

The Computer as a Design Analyst

Stress analysis of relatively simple components can, quite satisfactorily, be carried out using conventional engineering analyses. There are components, however, that are of such complex form that the classical strength of materials approach is simply inadequate and overdesign of components will be necessary to compensate for ignorance of operational stresses. In the aerospace industries, for example, this is clearly undesirable.

Finite element methods are now being used extensively as an aid to engineering analysis and design. The finite element method analyses complex components by breaking them down into small elements of simple shape. These elements are interconnected at their boundaries to form a mesh and the analysis is carried out by solving, mathematically, the stress in each element then combining these results to produce a stress analysis of the whole component. Finite element methods can also be used for solving thermal, vibration and many other engineering problems.

3-D modellers too, can be used as design analysis aids. Kinematic analysis of linkages and mechanisms is a common feature of 3-D modellers.

Chapter 2

CAD Hardware

2.1 INTRODUCTION

A computer-aided design (or manufacturing) system requires *hardware* (i.e. a computer and associated devices such as graphics terminals and plotters) and *software* (i.e. the programs of instructions which make the hardware operate in the desired manner).

This chapter describes the hardware components of a CAD/CAM system that could be used for a given design or manufacturing activity. The hardware elements are not related to any particular activity as the use of specific elements will become apparent in relevant chapters.

2.2 COMPUTER HARDWARE

Computers for CAD purposes can range in size from the large, general purpose, *mainframes* down to modern *micro* computers.

Computers are generally identified by their processor *word length*. The word length (i.e. 8 bits, 32 bits, etc.) represents the size of the data bus or the number of bits that can be transferred around the computer system in parallel. Generally speaking, the longer the word length the

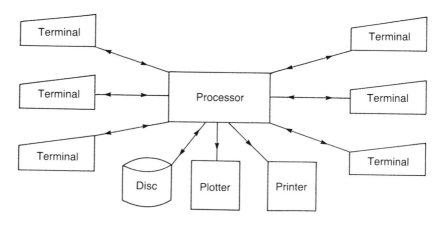

Fig. 2.1. Centralized computer configuration.

more complex the CAD system that can be operated. Until a few years ago, microcomputers had 8-bit, mini computers had 16-bit and main-frames had 32-bit word lengths. However, the trend is changing as 16-bit micros and 32-bit mini computers are now commonplace. This means that computer-aided design tasks which traditionally had to be carried out on expensive mainframes can now be run on cheaper mini- and microcomputers.

There are a number of basic ways in which the computer hardware can be configured for CAD use. Four methods will be identified.

(i) *Centralized system* — This approach is usually based on a main-frame computer which is capable of supporting a large number of terminals (20+) simultaneously, as shown in Fig. 2.1.

The mainframe would normally be dedicated to the CAD activity and is, in general, only used by large organizations. Following the initial capital costs, additional terminals may be added on at a relatively low cost. The danger is that if too many users require simultaneous access to the computer the response time may become irritatingly slow.

(ii) *Multiple host system* — With this configuration, several mini-computers each have a limited number of terminals connected to them. The mini computers are also interconnected so that the individual user appears to have the same facility as a centralized system.

Figure 2.2 shows a multiple host configuration.

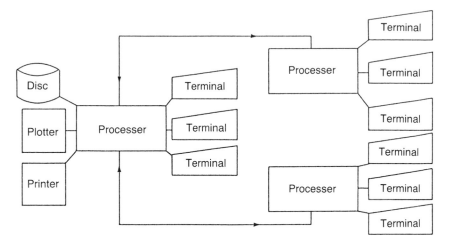

Fig. 2.2. Multiple host computer configuration.

Chapter 2

There is a distinct advantage of using this type of system in that if one computer were to go down then only a small number of terminals would be affected. However, the capital outlay for a multiple host system could be greater than for a centralized system.

(iii)   *Distributed processing* — A distributed system is one in which each terminal has its own computer. The computers are connected by a *network* arrangement to form an integrated system. The computers could be micros, minis or even mainframes, and users of the networked system can share common peripheral devices such as disc drives and plotters as shown in Fig. 2.3.

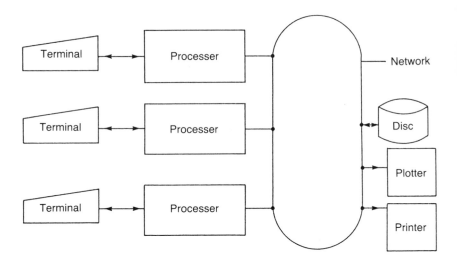

Fig. 2.3. Distributed processing system.

(iv)   *Stand-alone* — A stand alone system is one in which a user has access to a single *dedicated* computer. The computer drives a graphics screen and keyboard. The larger micros or small-sized minicomputers are used for this type of system. Stand-alone configurations are ideal for use by engineering design consultants, freelance engineers or colleges of education.

There are some disadvantages in using these systems, for example, the memory size is limited which may restrict the complexity of components that can be designed, and some stand-alone systems do not lend themselves to expansion.

Additional Computer Features

Main Memory

Together with a minimum 16-bit processor, the computer system must also have facilities for storing information. *Main memory* (working memory or immediate access store) is where all or part of the programs and data are held whilst in use. The memory capacity of CAD/CAM computers vary but typical sizes range from 256K bytes up to about 8 M bytes of RAM (Random Access Memory). Some mini and mainframe computers have a *virtual* memory facility where parts of programs and data are continually being moved (paged) between main memory and backing store. This feature enables the user to have virtually unlimited memory capacity. Modern main memory is *volatile* in nature (i.e. the contents are lost if the power is turned off) and a means of storing programs and data on a permanent basis is required.

Mass Storage

Magnetic disk storage is used as *backing* (or secondary) store for almost all CAD/CAM systems.

A variety of disk storage systems are available:

(i) Non-removable disk — such as the Winchester type where 4 M byte to 300 M byte of information can be retained. This type of disk is very popular for CAD applications as it proves most cost effective.

(ii) Removable disks — There are two main types of removable disk; hard disk and floppy disk. Hard disks are approximately 350 mm diameter and are only used on the larger minicomputer or main-frame based systems. Typical storage capacity of a single hard disk would be 20 M byte. Floppy (flexible) disks are used as mass storage media on microcomputer-based or small minicomputer-based systems. They range in size from 89 mm diameter to 203 mm diameter with storage capacities from 125 K bytes to 1.5 M bytes.

Magnetic tape can also be used as mass storage media and is particularly useful for archiving records or storing information over long-term periods.

Interfaces

All CAD/CAM systems require *interfaces* which link the computer with other elements of the workstation. An interface has two elements, a 'hardware' element, i.e. electrical components such as: P.C. Boards and ports to perform transmission and receipt of electrical signals, and 'software' elements to control the information and detect errors. There are two basic methods for interfacing the computer to peripherals, a *serial* connection using an RS232 or RS422 interfacing link, or a parallel connection for data interchange using a *parallel* interface.

There are two approaches to serial connection: *star* and *daisy chain*.

In a star system, information is passed between the computer and workstations (or terminals) by sending data down a single wire one bit at a time. The data are then converted into parallel information at the receiving end. Each workstation is attached to the computer individually by independent wires, hence the term 'serial star' connection. This is a low cost form of communication.

In a daisy chain system, the workstations are connected to the computer by one wire rather like a ring main. The workstations can be placed in priority order in the chain. Data is transmitted from the computer, down the wire, and is received by the appropriate workstation.

With a *parallel* interface system, information for one character is sent down a number of parallel wires simultaneously giving very high transmission speeds. There are some workstations which require this high speed of data transfer for their efficient operation.

2.3   WORKSTATIONS AND PLOTTERS

The engineer or designer needs to be able to communicate with the computer in order to input raw data and to gain access to the results of processing. A CAD system generally has one or more workstations for input (and some output) and a plotter for output.

2.4   WORKSTATION

The workstation is the engineer's vehicle for communication with the CAD system. Workstations comprise several items of equipment. The actual configuration of the equipment will vary from system to system, but the essential element of a workstation is a graphics display terminal. A graphics display terminal is a high resolution *visual display unit* (VDU) which normally comprises a cathode ray tube and associated circuitry plus a keyboard for instruction input. Some terminals include microprocesser intelligence for screen image manipulation and vector generation. Basically, three types of graphics display exist: Raster display, Vector display and Direct view storage tube display.

Graphics Displays

(a) *Raster Display*

This is a technique of building an image on a cathode ray tube and is the method currently adopted for displaying commercial television pictures. An electron beam is continuously cycled in a scanning pattern of adjacent horizontal lines across the face of the CRT. To create an image or outline, the intensity of the electron beam is varied for all points which form part of the image. To position a spot on the cathode ray tube, horizontal and vertical deflection voltages are applied to the beam in proportion to the deflection. The electron beam striking the rear of the screen causes the phosphor coating at that point to glow for a finite duration. For the image to be maintained, it must be re-traced before the eye can detect any decay in the glow. If the retracing frequency falls below about 40 pictures/second, then the image will flicker. In some instances, a small computer is used in addition to the main computer in order to *refresh* the display.

When straight lines or characters are to be displayed in a refreshed system, special hardware facilities are often used to provide a fast response. A straight line is produced by generating successive spots along the line and characters are generated from standard sub-routines. The screen is broken up into picture elements (or *pixels*) which are illuminated as necessary to form the picture.

Figure 2.4 shows a raster type graphics display.

(b) *Vector Display*

A vector display uses a cathode ray tube with a phosphor coating on the inner surface of the display screen. A beam of electrons is continuously directed onto the phosphor coating, drawing the desired picture line by line, over and over again. Data, which describes the end points of each line, is continuously being fetched from memory so that the entire picture image can be re-drawn (refreshed). Vector refresh displays employ very fast acting circuitry so that operations such as *zooming* (enlarging a specified area of a picture to examine fine detail) can be carried out very quickly. This type of display offers the user the ability to operate directly on the viewed object, making changes in memory

Fig. 2.4. RASTER type graphics display (courtesy Intergraph).

which are immediately reflected on the screen. In this way real-time modelling and screen image manipulation (such as: rotating an object on the screen so it can be viewed from another angle) are easily effected.

Light pens are very often used in conjunction with a vector refresh display, and this type of display has an outstanding ability to selectively erase any part of the displayed image.

Figure 2.5 shows a vector refresh graphics display.

(c) *Storage Tube (direct view)*

The direct view storage tube is a version of the cathode ray tube in which the coating on the viewing face retains, for a period of time any image generated by the passage of an electron beam across its surface. The DVST is one solution to the problem of the cathode ray tubes fleeting image and of the necessity to refresh the display to avoid flicker.

The DVST is constructed like a conventional electrostatically-controlled cathode ray tube but with an additional grid electrode and a second electron gun. The additional electron gun provides a broad low

Fig. 2.5. VECTOR refresh graphics display with light pen in use (courtesy V.G. Systems Limited).

energy beam of electrons and is called a *flood gun*. The grid electrode is normally formed of fine wires spaced as closely as 250 per 25 mm and is placed near the tube viewing face. The surface of this grid remote from the viewing face, is coated with a film of good insulator. The operation of the storage tube depends on the phenomena of secondary emission due to electron bombardment. The grid electrode is negatively charged by the application of a suitable voltage before writing. When the main write beam of electrons is moved so as to display a line, the grid becomes positively charged in the area where the tightly focused beam strikes it, thus allowing electrons from the flood gun to be accelerated through onto the screen, causing the phosphor to glow. The image is displayed continuously, without the need for any retracing of the write beam. The image can be erased from the screen by negatively recharging the grid. Although the direct view storage tube overcomes the need to refresh the display, and hence saves on computing power, the principal disadvantages of the DVST are:

(i) That it cannot be selectively erased, that is, parts of the image cannot be individually detected, and

(ii) that a light pen cannot be used, since continuous refresh is not available.

Figure 2.6 shows a graphics display of the direct view storage tube

Fig. 2.6. Direct View Storage Tube graphics display (courtesy Tektronix U.K. Limited).

type.

A graphics display can be used in conjunction with other devices such as joy sticks and digitizing tablets for input of data to the system. It is also the principal medium for displaying drawings.

2.5 DIGITIZING BOARD

The digitizing board is a common method of data input in a computer aided draughting sytem. The device allows the direct input to a compu-

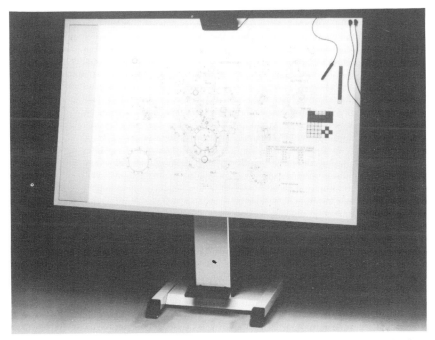

Fig. 2.7. High resolution AO digitizing board (courtesy Benson Electronics Limited).

ter, of data from an outline drawing in the form of *digital co-ordinates*. The digitizer board looks similar to a conventional drawing board and functions as a fine measurement grid. Positions on the board are indicated either by stylus (pen) or by cursor, depending upon the speed, accuracy and type of work being carried out.

The stylus or cursor can be moved over the surface of the board, and will contain a switch which enables the user to register x, y co-ordinates at any desired position. These co-ordinates are fed directly to the computer or to an off-line storage device. There are many digitizers on the market which operate on various principles. Figure 2.7 shows a digitizing board.

2.6 GRAPHICS TABLET

The tablet is simply a small, low-resolution digitizing board and is often used in conjunction with a graphics terminal. The surface of the tablet corresponds to the face of the CRT. The user may point to areas of the flat surface of the tablet with a stylus and the position of the selected area may be transmitted to the computer and the position then echoed

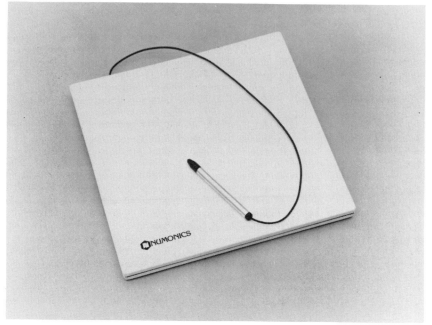

Fig. 2.8. Graphics tablet (courtesy Comtec Data Systems Limited).

on the screen. The tablet is particularly suitable for interactive design since it allows the engineering designer to work naturally with stylus in hand (simulating free hand sketching).

The motion of the stylus is replicated on the screen allowing the designer to check his/her actions. The computer can straighten lines and present a clean picture if it is fed with accurate salient points. Figure 2.8 shows a graphics tablet.

Often, the tablet is divided into areas with significant meaning to the draughting system. This is called a *menu* or function block. In effect, the menu gives access to a specific instruction or small drawing routine in the computer system. The use of a menu eliminates the need for typing coded instructions and simplifies the CAD process. Figure 2.9 shows a typical menu for a computer-aided draughting system. Various types of tablet exist, but most are based on the RAND design. The RAND tablet consists of a matrix of crossed conductors which has been fabricated by printed-circuit methods. The circuitry uses switching techniques to apply pulses to the conductors in sequence, thus digitally coding their individual locations. When a stylus is touched to the surface, it picks up pulses capacitively from the closest conductors, which specify its location.

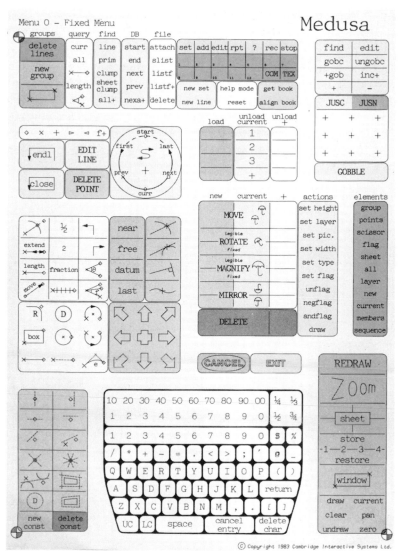

Fig. 2.9. Command menu for CAD draughting system; this menu is part of a set (courtesy Cambridge Interactive Systems Limited).

2.7 JOYSTICKS, TRACKING BALLS AND THE TRACKING MOUSE

A command menu instructs the computer what to do, but it does not always specify where on the display to execute the instruction. To specify this position, a joystick, tracking ball or mouse can be used instead of a tablet to move a cross (cursor) to any position on the face of the display screen.

A *joystick* is an analogue, lever-operated, two-axis control device, similar to the controls of an aircraft. Two potentiometers are attaached to the joystick, one moving in proportion to the up–down motion and the other in proportion to the left–right motion of the stick. Voltages from these potentiometers are converted to digital form and are input to the display. Since motions of the joystick may be considered as an *xy* co-ordinate reference frame, the cursor can be moved directly by the joystick. A trigger switch can be placed on the joystick for inputting on/off commands, and versions are available which can be rotated as well as moved in the *X* and *Y* directions. A joystick device is shown in Fig. 2.10.

Tracking Ball

An alternative to the joystick is a tracking ball. The tracking ball comprises a sphere, approximately 125 mm in diameter, mounted on

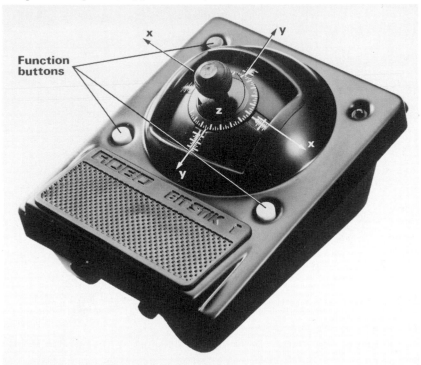

Fig. 2.10. Joystick with zoom control (courtesy Robocom Limited).

rollers in a recessed space so that its exposed top is essentially flush with a table surface. The ball can be rolled with the palm of the hand, causing potentiometers attached to the rollers to turn. Thus, it is possible to roll the cursor around the screen.

Tracking Mouse

A tracking mouse is a small enclosed analogue device about 100 mm in length. It has a ball or two sets of small wheels on its underside which rotate when the device is pushed along a surface. As the ball or wheels rotate they generate co-ordinate data which is transmitted to the work-station, via a fine wire, to determine the cursor position on the graphics screen.

The majority of these devices incorporate function buttons for menu select commands and so on. The mouse can be operated on any flat surface, and does not require a tablet as would a puck or electronic pen. One drawback with the mouse is that if it is lifted from the surface its origin is lost and therefore cannot be reliably used as a menu select pointer.

An analogue mouse is shown in Fig. 2.11.

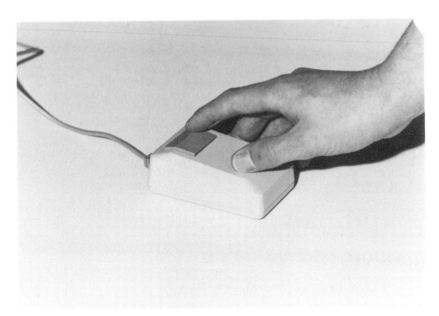

Fig. 2.11. Analogue mouse (courtesy Apple computers).

2.8 LIGHT PEN

The light pen is a small rod-shaped device, rather like a felt-tip pen in shape and size, which is connected to the graphics terminal. The pen can be used as an input device for drawing or moving items on the graphics screen. Inside the barrel of the pen, there is a photosensitive fibre optic bundle. An aperture at the tip of the pen allows light emitted from the screen to be passed down the fibre optic cable to a photomultiplier, which in turn passes a signal back to the computer. Thus it works in reverse to a normal pen.

Once the computer has detected the spot at which the pen is pointing, it usually generates a small pattern onto the screen, such as a cross or circle, to inform the designer that the pointing activity of the pen has been detected.

Figure 2.5 shows a light pen in use.

In order to draw images onto the screen or to move images around, additional commands such as DRAW or MOVE must be supplied and may refer to drawing/moving spots or vectors or lines onto or around the screen.

To support the activity of light pens and input devices, mentioned earlier, to a graphics terminal, a considerable amount of software must be available.

2.9 TERMINAL

Finally, an 'alphanumeric display terminal', complete with keyboard, is an essential part of the workstation, displaying prompts and queries which aid the designer. It is also extremely useful when editing instructions to drawings for example, to increase the precision of dimensions input with outline data using a digitizing board. The keyboard is mainly used for text and, of course, is the only means of communicating with the computer when operating outside the CAD mode. The use of two screens in a workstation is useful for separating command text or 2-D drawing data from pictorial output.

Figure 2.12 shows a workstation with display terminals, keyboard, graphics tablet and pucks.

2.10 GRAPH PLOTTERS

The plotter is the most common hard copy (permanent) output device used in computer-aided design systems and is the peripheral which produces the finished drawing. There are six basic types of plotter

Fig. 2.12. Complete CAD workstation (courtesy Intergraph (Great Britain) Limited).

available. One of the most common is the drum plotter. *Drum plotters* have paper partially wrapped around a drum which rotates giving longitudinal paper movement under a pen carriage. The pen carriage traverses parallel to the axis of the drum producing lateral pan movement. The paper is in roll form, usually 40 metres in length, which can be plain or pre-printed, with sprocket holes along each edge. Sprockets on the drum maintains paper alignment.

Figure 2.13 shows a typical drum plotter.

Operation of the plotter relies upon movement of the drum and pen carriage which can be simultaneous. Movements take place in single discrete steps (0.1 mm to 0.25 mm).

There are six operating modes.

(1) Drum rotation forwards/backwards.
(2) Carriage movement left/right.
(3) Pen up/down.

Fig. 2.13. Single sheet 8-pen plotter (courtesy Benson Electronics).

The drum and carriage motions give the plotting axes, x and y. Movements at 45° to the x and y axes can be obtained by simultaneous movement of drum and pen carriage; this allows eight basic step directions as shown in Fig. 2.14.

Lines which do not lie in one of the eight basic plotting directions must be approximated as illustrated in Fig. 2.15.

A line drawn using this method appears straight to the human eye as the incremental steps are small. The engineer or designer does not have to consider this technique as the software is usually provided by the plotter manufacturer. Standard routines also cater for curves, symbols, chain lines, etc.

Flatbed plotters are used when increased drawing accuracy is desirable. As the name implies, flatbed plotters operate by plotting on paper which rests on a flatbed. The paper is held down by vacuum or other means. The pen moves vertically and horizontally across the bed. One directional movement is supplied by a gantry which straddles the bed. The gantry is driven by d.c. servo motors and may include optical encoding equipment for positioning. The other direction is facilitated by a pen turret which traverses the length of the gantry.

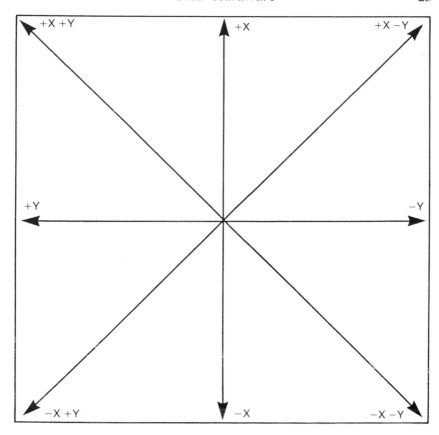

Fig. 2.14. Eight basic step directions of a drum plotter.

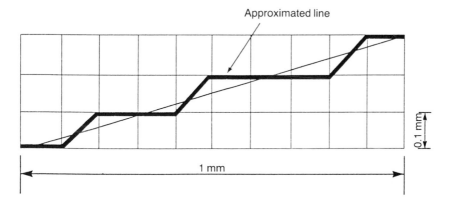

Fig. 2.15. Line approximation for lines which do not lie in one of eight basic directions.

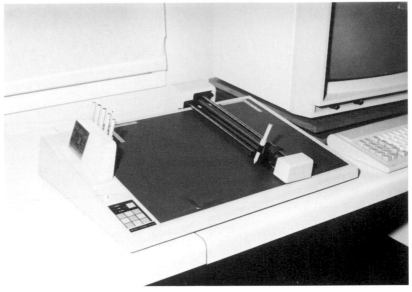

Fig. 2.16. Multi pen flatbed plotter (courtesy Denford Machine Tools Limited and Watanabe).

Figure 2.16 shows a typical flatbed plotter.

Some flatbed plotters can etch plastic or metal to form master plates for creating manufactured components. Plotting beds can be quite large, in fact, those used to assist aero design may be as large as 6×15 m. *Vertical bed plotters*, or easels, are an alternative to the flatbed type. This type of plotter is like a flatbed rotated through 90° so that it assumes the position of a conventional drawing board. Flatbed plotters combine high speed with accuracy.

Pens

Many plotters are designed so that virtually any pen can be used in conjunction with it. Such an arrangement gives the designer almost infinite control over parameters like the colour of ink being used and the size and type of nib.

Some plotters allow a number of pens to be selected by the machine at any one time. This can be achieved by using a carousel which holds several pens in the same unit as the one that is actually drawing. An alternative is to site the pens somewhere away from the main plotting

Fig. 2.17. Electrostatic printer/plotter (40,000 dots per square inch) (courtesy Versatec Electronics Limited).

area and let the plotter select from a rack. Felt tip pens tend to be the first choice of plotter manufacturers. Their advantage is that they offer a wide range of colours and tips. Generally, the higher the resolution of the plotter, the finer the tip of the pen which is used.

Photoplotters include a light source in the head for producing an image on photosensitive media and are particularly suitable for the production of printed circuit photographic masters. Other applications include the production of microfilm record cards of drawings; mechanism motion animation films; integrated circuit master; and graticule and Moiré fringe masters.

The fifth type of plotter to be discussed is the *electrostatic printer/plotter*.

Figure 2.17 shows an electrostatic machine which consists of an electronic dot matrix which can print dots on to charge-sensitive paper, the paper being fed round rollers and over the matrix. The drawing is produced in raster format as the paper passes over the writing head. Dot density is of the order of 40,000 per square inch. The drawing sheet with its electrostatic image is then fixed by passing the paper through a cloud of ink powder and finally curing with heat. Plotting speeds are very fast and paper widths of 1.75 m are available. Some electrostatic machines produce colour plots — the principle of operation is shown in Fig. 2.18.

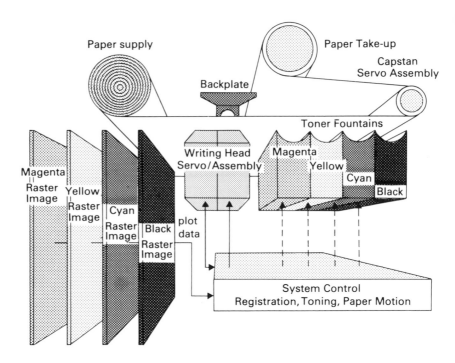

Fig. 2.18. Principles of operation of electrostatic colour plotter (reproduced by kind permission of Versatec Electronics Limited).

Ink Jet Plotters

An ink jet plotter is a raster device which can produce full colour, shaded pictures.

A sheet of paper is wrapped around a drum and secured in place. The drum rotates about a horizontal axis at approximately 260 rev/min. A drive mechanism then carries three ink jets on a carriage from one end of the drum to the other. As the jets move past each column on the paper the control mechanism allows a metered volume of ink to be fired at the paper from the ink jets. The position of the drum and combinations of the fired ink form an image on the paper which corresponds to a bit map in memory.

Typical resolution varies between 100 and 200 spots per 25 mm and paper sizes up to A0 can be used.

Chapter 3

CAD Software

3.1. INTRODUCTION

At the heart of any CAD system is its *software*. CAD systems rely upon software for their efficient operation. Since the early 1960s there has been an ever-increasing development and consolidation of computer-aided design software which has provided industry with the ability to integrate computer-aided design and manufacturing systems. This chapter discusses the essential software elements and points out some of the factors to be taken into consideration when selecting a computer-aided design package.

3.2 TYPES OF SOFTWARE

For a CAD system to function correctly, two distinct types of software are required.

(i) *Systems software* — including an operating system to control the operation of the computer and, associated systems software such as editors and filing systems, and

(ii) *Applications software* — This is the specific computer-aided design package.

Systems Software

The *operating system* is the major system software element in a computer-aided design system. Its main purpose is to make the computer hardware function effectively. Modern operating systems are made up of a collection of integrated program modules which collectively serve various functions such as: hardware management, user interfacing, and the provision of user facilities. The operating system, as well as managing the internal working of the computer also controls the peripheral devices.

Most CAD systems have some form of auxiliary storage device, usually a disk drive, and one section of the operating system has to oversee the disk storage system. This part of the operating system is called the *disk operating system* (DOS) and it has the task of organizing the transfer of files between disk and memory as well as keeping a check on the location of files.

The operating systems software also acts as an interface between the user and the computer. This interface enables the user to communicate with the computer via the operating system language. A common operating system for the larger CAD systems is the *UNIX* system, which is a product of Bell Laboratories, whereas the smallest CAD systems usually run under *MSDOS* (microsystems disc operating system) or *CP/M* (control program for microcomputers) based systems.

Utilities

Systems software for carrying out specific tasks in CAD, such as editing (drawings or text) or filing (drawings, parts descriptions or text) are also provided by manufacturers and are called *utility software*. However, editing software can also be included in the applications package.

Editor Utility

When a drawing is being produced on a CAD system mistakes may be made or the user may wish to alter the shape of a part or re-dimension a whole drawing. In other words the user needs to have flexibility and this can be provided by the editor.

Essentially, editing involves adding new information, deleting unwanted information, or correcting existing information. The data can then be written to disk in its new form.

Drawing — editors usually allow addition, modification or erasure of lines, points, circles and arcs via the screen.

Filing Systems

When a drawing has been produced or when text has been entered into the system, the information has to be filed away for later retrieval. Filing utility programs are usually provided to ensure that information is filed in the most efficient manner to maximize storage capacity and provide ease of retrieval. Drawing information is usually held in drawing files which are structured to maximize the speed of finding selected objects at specified locations within a given area. It is usual that all graphical records have pointers that may be used to provide access to additional records containing non-graphical data. In this way the drawing file forms the hub of a multipurpose data base, which can provide quantities, dimensions or other properties required by the user.

Drawing files are held on disk, as random access files, each capable of holding a large number of drawings. A typical drawing requires between 0.1 and 0.2 M byte of store, so in the region of 300 drawings can be held on disk. Drawing files may be copied to other discs for archiving purposes.

Applications Software

It is the applications software (*package*) which turns the CAD hardware into a useful system. There are a number of different types of computer-aided design program, but generally speaking they fall into three main categories: Computer-aided draughting, Geometric modelling, and Finite Element Analysis systems.

Computer-Aided Draughting Software

Most draughting systems can be bought off-the-shelf and offer two-dimensional draughting and plotting facilities as a minimum requirement. Hard copy drawings and charts can be produced following the initial design work created interactively by the draughtsman. The more sophisticated software packages allow parametric symbol creation and drawing orientation, and some packages enable the drawing data to be used in the manufacturing process.

Computer-aided draughting systems are having a significant impact on drawing office operation.

Geometric Modelling Software

It is possible for a designer to use a computer system for modelling engineering components. The problem arises, however, that mechanical engineering components are three-dimensional objects and are sometimes complex in form. It is difficult to represent a 3-D shape on a two-dimensional graphics screen and the software to perform this task is necessarily very complex. Geometric modelling systems are very useful tools as they give the user a real-life visualization of the product under consideration. The simplest type of modelling system displays components as stick-like figures but the more sophisticated packages can model and display the component as a solid. Additional features such as hidden line removal, colouring and selective shading are possible. Data from a geometric model definition can be used by other related elements of the CAD/CAM system, such as finite element analysis, draughting or manufacture. One drawback with a modelling package is that to describe the geometry of an object in 3-D could use up to eight times the disk storage space as it would to hold the same component information from a 2-D drawing.

Finite Element Software

Finite element methods are used as an engineering analysis and design aid. There is an enormous amount of software available for the analysis of specific engineering components and most packages now have sophisticated pre- and post-processors for automatic model definition (meshing) and graphical visualization of results. Traditionally, finite element software was used only on mainframe and powerful mini-computers but, more recently, packages are being announced which can operate on desk-top machines.

Draughting, modelling, and finite element systems are usually written in machine-independent, high-level languages such as *ANSI FORTRAN* or *PASCAL*.

Method of Use

CAD systems (particularly draughting and modelling) can be used in one of two ways or combinations of both. That is *menu driven* software or *command driven* software. In a menu-driven system the designer is guided through the package by being offered a menu of options. Selection can then be made by a variety of methods (light pen, cursor, joystick etc.). Menus are hierarchically structured and are often used to provide access to a particular facility within the main menu. The menu can be etched on to a digitizing tablet, overlayed onto a tablet, or can appear on the screen, depending upon the system. The major benefit of this menu approach to CAD is that a user can be lead systematically through a pre-determined, logical sequence of events to reach a design goal. Such a system is therefore ideal for newcomers to CAD.

With a command-driven system the designer creates the geometrical information by entering individual command lines. This method gives the designer greater freedom in the way in which the geometry is developed and, as a result, more complex shapes can be produced. However, the use of such a system is not as easy as a menu-driven type and is therefore only used by people who are familiar with a particular system.

Databases

For an engineering company with any computer-aided design equipment, a *database* is a desirable tool, particularly if the intention is to eventually integrate the total CAD/CAM activities.

A database is a generalized integrated collection of data which is structured on natural data relationships so that it provides all necessary access paths to each unit of data in order to fulfil the differing needs of all users of the system. In a database system the data is stored once only. The different application programs (such as modelling, draughting, numerical control) can access the same database via a set of controlling programs known as the *database management system.*

Figure 3.1 shows how both design and manufacturing activities can share a common database. A database program system is usually large and requires considerable computer resources. It is important, therefore, that the design of any comprehensive CAD/CAM database is considered with care.

3.3   SELECTING THE SOFTWARE

There now exists a large number of computer-aided design and manufacturing software packages ranging from simple 2-D draughting systems to total integrated CAD/CAM suites (encompassing design, analysis and manufacture). It is of vital importance to an engineering company to specify and select the right package to satisfy their requirements.

There are a number of factors which have to be taken into consideration when selecting computer-aided design software, and these include:

  (i) *Supplier* — It is important to establish the reputation of the CAD/CAM software supplier. Questions which ought to be asked include: Who is the supplier? What is the suppliers standing? Is a trial period allowed? What would be the delivery time? and, What are the contract terms?

 (ii) *The package itself* — Most packages tend to be written in a general purpose manner and whilst this is fine from a manufacturers standpoint, it could conflict with established principles and methods of operation within an engineering company. When

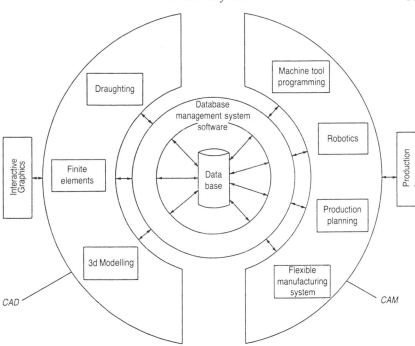

Fig. 3.1. Computer Aided Design and Manufacturing activities sharing a common database

acquiring any CAD/CAM software it is important therefore, to enquire about the possibilities of own-coding and ease of tailoring the system to meet the company specification. It is important to know when the package was released and also to have knowledge of any recent enhancements.

(iii) *Training and support* — Most CAD software suppliers offer some form of training for their users. It is desirable to know what type of formal training is offered and to what extent a user is competent after an introductory course. Support can come in many forms and it should be established what degree of support is provided. Aspects to be considered include the response time for maintenance requests, whether a regular newsletter is published or if a user group is established.

(iv) *Ease of use* — User friendliness is a major aspect of software packages. The software, whether for draughting, modelling or finite elements should be easy to use in order for the engineer to obtain maximum benefit from the system. If the package runs interactively a HELP facility is often provided.

If the computer hardware supports a number of different user packages a desirable feature is that the user can gain access to a specific CAD package, without the need to have knowledge of the computer or operating system.

(v) *Technical details* — If acquiring a CAD system and a computer at the same time the selection problems are not so acute as the hardware and software can be chosen to be mutually compatible. However, if a CAD package is required to run on an already installed machine, then there are certain aspects to consider: for example, what is the minimum memory requirement, what is the minimum hardware configuration? Other general technical considerations might include: The portability of the package, the ease of interfacing, the language the software is written in, and the user documentation supplied.

(vi) *Security* — It is likely that drawings, models and results of analyses produced on a CAD system will be of a technically confidential nature. It is essential, therefore, that only authorized persons be allowed access to the workstation.

Security log-on routines and individual user passwords can go some way toward ensuring security of company information, and it is desirable that such features can be provided by the CAD system supplier.

(vii) *Procurement* — CAD systems including hardware and software can be purchased outright, hired or leased. The choice depending upon the requirements of the user. The cost of the system will be a major factor in the consideration of the acquisition of a CAD system. There are software directories available listing the application, the supplier, operating system details and user application software for a host of computer-aided design and manufacturing systems.

Chapter 4

Computer-Aided Draughting

4.1 INTRODUCTION

The purpose of a draughting system is to automate, where possible, the production of engineering drawings. Draughting is one of the few aspects of the design process which can easily gain considerable benefit from automation, and for this reason has been the focus of attention for Computer-Aided Design system vendors and buyers alike. Very often, the production of a detailed drawing can create hold ups in manufacture and an increase in draughting productivity could well be a valid reason for the installation of a CAD draughting system.

Computer-aided draughting is sometimes regarded as the same as computer-aided design, sometimes as part of it. CAD can, quite appropriately, denote either.

This chapter describes the salient differences between conventional and computer-aided draughting. The essential features of computer-aided draughting are discussed and a case study of a 2-D draughting package is presented to show how a drawing may be constructed interactively. Mention is made of geometry visualization systems providing three-dimensional views of 2-D drawings and, finally, the major advantages and limitations of a computer-aided draughting system are given.

4.2 DRAUGHTING

The construction of a drawing may be achieved using one or both of two basic techniques, conventional draughting and computer-aided draughting.

Conventional draughting: This involves a draughtsman operating a standard drawing board utilizing drawing instruments and paper (or film). The major attractions of this conventional approach to draughting are low capital expenditure and high system flexibility. Unfortunately, however, this high degree of flexibility also applies to the quality of the final drawings, in terms of appearance, consistency and accuracy. Indeed, the whole accuracy of the drawing is a function of the manual dimensioning of the drawing. The dimensions are not inherently related to the drawn outline, and, even if a profile is drawn correctly, there is no guarantee that it will be correctly dimensioned. This situation provides endless potential for errors.

Furthermore, the linking of data from a manually produced drawing with a manufacturing activity such as N.C. tape preparation is tedious, time consuming and error prone.

Computer aided draughting: In contrast to conventional draughting, computer-aided draughting systems can be highly capital intensive requiring the need to purchase or hire a computer, work station and software.

However, they do provide many potential benefits including: improved quality of design information, reduction of draughting errors, better staff utilization and improved draughting productivity. Draughting systems are probably the most easily used and most beneficial elements of a CAD system.

Computer-aided draughting systems have been in existence since the early 1960s. At that time, they required large, powerful and expensive computers and were restricted to use in large industries such as automotive or aerospace. Graphics capabilities were poor and data entry was via the alphanumeric keyboard. With the development of graphics displays and the introduction of mini-computers in the early 1970s, computer-aided draughting was used in more diverse engineering industries. Printed circuit board design was one area which gained enormous benefits from such systems. Now in the mid 1980s there are over 100 companies offering two-dimensional computer draughting, ranging from low-cost single-user systems to high-cost multi-user integrated design systems.

In the modern computer-aided draughting system the drawings are

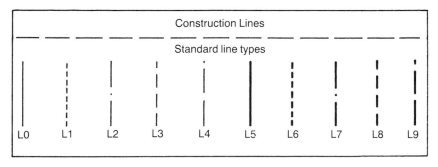

Fig. 4.1. Common linetypes used in Computer-Aided Draughting.

constructed interactively by the user supplying information to a computer via commands or with a menu-driven series of instructions. The drawings can easily be edited, manipulated and stored on backing store for future retrieval. A plotter is used to produce the final drawing to high standards of accuracy and line consistency.

One major advantage of using computer-aided draughting methods to produce drawings is that the link with manufacture is easily effected.

4.3 FEATURES OF COMPUTER-AIDED DRAUGHTING SYSTEMS

Typical computer-aided draughting systems employ a computer, workstation (including the elements discussed in Chapter 2), and a software package. Most draughting systems have a good user interface making them easy to use, the majority of packages being menu driven.

Two-Dimensional Draughting

The basic elements comprising any drawing outline are points and lines, if arcs, circles and conic curves are considered to be specific examples. A sheet of paper, or indeed the surface of a display screen are both examples of *two-dimensional* surfaces. Therefore, all drawings whether orthographic or pictorial have to be created in terms of 2-D drawing elements. It is possible to define points, lines, arcs and circles in a variety of ways in order to build up a drawing. Figure 4.1 shows a set of possible line types, whilst Fig. 4.2 shows typical circle and arc functions.

Drawing Procedure

The procedure for drawing a component may be as follows:

The image on the screen is drawn on a number of independent *conceptual layers*, often as many as 99 are available for use. For example, the construction lines might be stored on one layer and the dimensions on another.

Each layer can be thought of as being a transparency which can be overlayed on top of others in order to build up the image of the drawing. This layering technique enables the computer to store the drawing in a structured manner, allowing the user to store, modify, edit, etc., certain portions of a drawing without affecting the rest.

The user lays out the construction lines on one selected layer. The component may then be drawn on remaining layers using options from

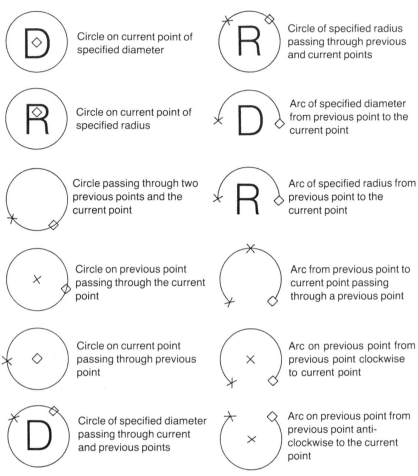

Circle on current point of specified diameter

Circle of specified radius passing through previous and current points

Circle on current point of specified radius

Arc of specified diameter from previous point to the current point

Circle passing through two previous points and the current point

Arc of specified radius from previous point to the current point

Circle on previous point passing through the current point

Arc from previous point to current point passing through a previous point

Circle on current point passing through previous point

Arc on previous point from previous point clockwise to current point

Circle of specified diameter passing through current and previous points

Arc on previous point from previous point anti-clockwise to the current point

Fig. 4.2. CIRCLE and ARC functions (reproduced by kind permission of Cambridge Interactive Systems Limited).

the menu. With a minimum of practice this can be done faster than by conventional means, since portions of the drawing may be translated, copied, rotated, and mirrored either singly or in some user defined groups. Figure 4.3 shows how simple definitions may be repeated.

Fillets, chamfers and hatched sections can quickly and easily be added. Hatching lines might allow user defined sizes and angles as shown in Fig. 4.4. Conventional representations of threads, knurling, welds, etc., may be added using standard options. These representations are stored as symbols and can be recalled from a library when required.

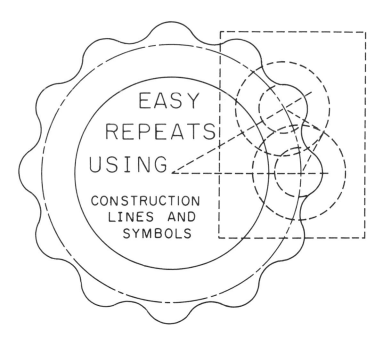

Fig. 4.3. Diagram to show how repeat features can be easily effected (courtesy Micro Aided Engineering).`

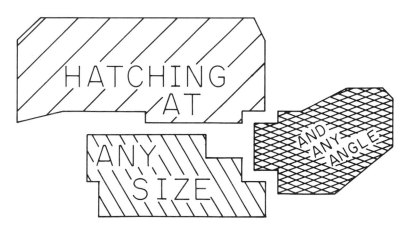

Fig. 4.4. Hatching facilities in Computer-Aided Draughting (courtesy Micro Aided Engineering).

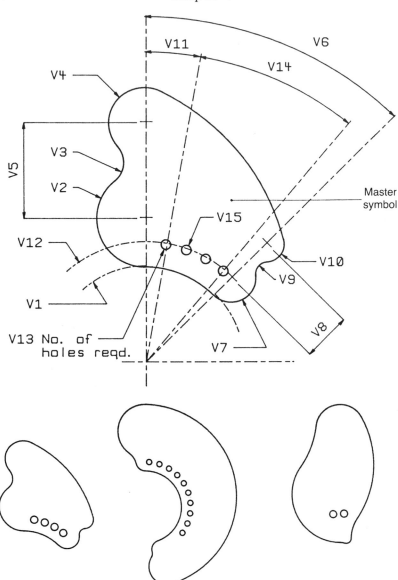

Three symbols generated from the PARAMETRIC master.

Fig. 4.5. Parametric symbol (courtesy Micro Aided Engineering).

Parametric symbols may also be drawn by first defining them in the system and recalling them when required.

If a component is manufactured in a number of different variations on a basic theme, it is possible to construct one symbol to produce all members of that family of components. The parametric symbol is constructed in the same way as a standard symbol except that features (salient dimensions) which must be altered are assigned variable names in place of constant values. These variables are then given values when the user calls up the symbol for use in the drawing (Fig. 4.5).

Parametric pre-processing is a very powerful CAD tool, enabling standard engineering shapes, e.g. shafts, bearings, seals, etc., to be drawn simply by inputting certain key sizes.

On completion of the drawing, dimension lines and toleranced dimensions may be added. The ability to generate vertical, horizontal and angular dimensions to a standard form is a prime requisite, most draughting packages therefore provide automatic dimensioning routines. See Fig. 4.6.

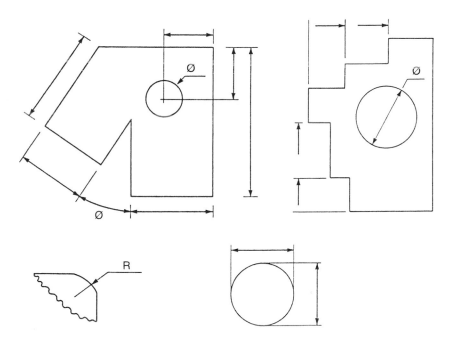

Fig. 4.6. Automatic dimensioning feature.

A *windowing* facility can be used to fill the screen with a specified portion of the drawing detail. This is a viewing facility that could be compared to having a powerful magnifying glass to examine fine detail. Zooming in on drawing detail does not modify the drawing in any way.

Figure 4.7 shows a windowed section of a drawing to reveal fine detail.

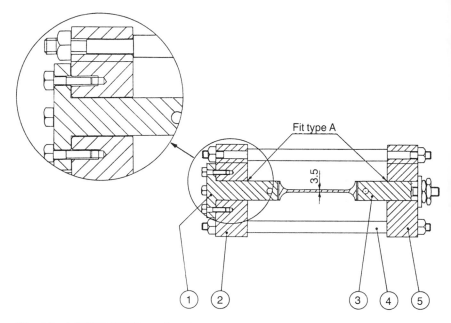

Fig. 4.7. A WINDOWED section of a drawing to reveal fine detail.

Finally, text in various fonts and sizes, standard notes and other items such as company logos, may be added to the drawing.

In order to save time during the creation of a drawing, most CAD systems have default settings for the various commands. These settings can be controlled by the user and enable the system to be instructed to use given values in a number of different situations — for example to use always a 0.2 mm chain line for centre lines, unless told otherwise. This feature can be extended to such items as arrow heads and text sizes and goes some way towards ensuring that pre-defined company standards are used automatically.

If, during the production of a drawing, the user wishes to erase a line or change some text, a DELETE command is usually available which is used to electronicaly erase drawing detail. This is one activity where the computer scores over the draughtsman. Unlike pencil and paper draughting, where line or text erasure often mars the quality of the finished product the CAD system shows no trace of change as only the final drawing will be plotted out.

The user can specify which items are to be deleted by selecting the DELETE mode and then pointing to them with the cursor. Most systems have built in safety features so that the user has to give secondary confirmation of the items to be deleted.

Drawings may be plotted or *archived* on backing store so that they may be retrieved at a later date. The created drawing is easily recalled and modified to accommodate design changes, most systems being provided with a series of editors for the deletion, replacement or modification of individual drawing elements.

When a drawing requires modification, the draughtsman retrieves the drawing by typing in a drawing number. The computer system fetches a copy of the drawing from disk and places it in the drawing area in the computer's main store. From this point on the user is only ever operating on a copy of the master drawing and therefore cannot corrupt the master. When drawing modification is complete, the revised drawing can be stored away as a separate new drawing or can be made to overwrite the original master drawing on disk.

4.4 DRAUGHTING CASE STUDY

In this section, the features included in a typical microcomputer-based draughting package will be described. The description is not intended as a comprehensive case study of a computer-aided draughting system; instead it provides an outline of the individual facilities available, which, when combined together form a system capable of producing a fully-dimensioned, working, engineering drawing.

The System

The system to be described is the *MAEDOS* two-dimensional draughting system, which is a product of Micro Aided Engineering Limited. The *MAEDOS* draughting package is part of an integrated, modular family of design, draughting and manufacturing aids, which has a common database allowing a very high degree of flexibility in any particular application. Figure 4.8 shows how the draughting element forms part of the overall CAD/CAM system.

MAEDOS software is implemented on a 16-bit microcomputer, running under UNIX- or MSDOS-based operating systems.

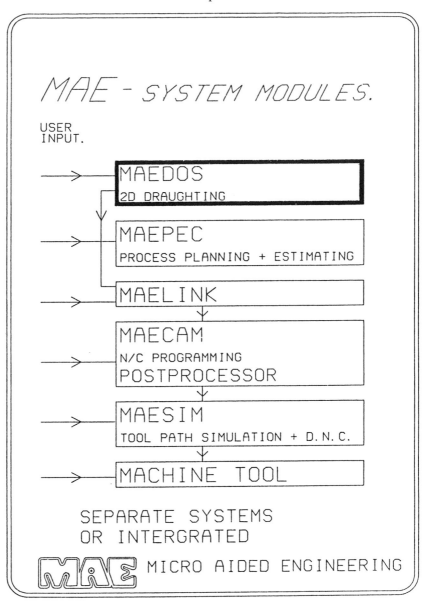

Fig. 4.8. Diagram to show how the draughting element forms part of the integrated system (courtesy Micro Aided Engineering).

The basic system supports two users. However, this can be expanded to support additional workstations by adding extra memory and increas-

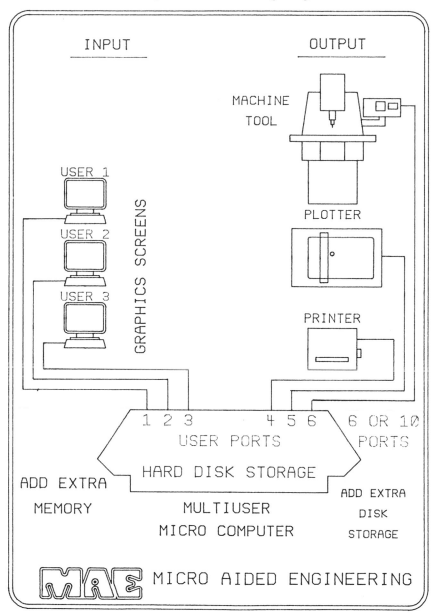

Fig. 4.9. Typical CAD/CAM user configuration (courtesy Micro Aided Engineering).

ing magnetic disk storage capacity. A typical configuration is shown in Fig. 4.9.

The workstation layout is designed to provide an ergonomically efficient working environment for the user and consists of the following:

A FRONTIER 19″ colour graphics display screen, a joystick which moves the cursor to any position on the face of the display screen, an alphanumeric keyboard for data and command input, and a plotter. See Fig. 4.10. A digitizing tablet and printer may also be connected to the system if required. However, for ease of use, the menu can be made to appear on the screen and features selected by positioning the cursor on the required item, using the joystick, and pressing the space bar on the keyboard to execute the command.

Fig. 4.10. Workstation showing disk drive, joystick, keyboard, display screen and plotter.

The following procedure describes how the *MAEDOS* system is used to produce a two-dimensional drawing of the component shown in Fig. 4.11.

Stages in Draughting

(1) Switch on machine and load the *MAEDOS* software into memory from the hard disk. This results in a question and answer session between user and computer to determine preliminary information such as: units to be used for drawing (inches, mm or cm), scale of drawing, and drawing sheet size.

(2) The menu then appears on the screen. A toggle switch enables the

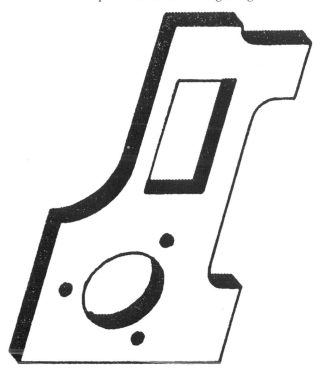

Fig. 4.11. Isometric view of component to be drawn.

menu to be switched on or off.

Figure 4.12 shows the menu. The menu can be logically divided into five areas:

(i) Point mode commands — e.g. NEAR will automatically select the closest point to the cursor, for example, end of a line or centre of a circle.

(ii) Function commands — e.g. DRAW will draw a line from the

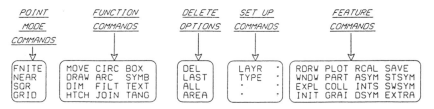

Fig. 4.12. Draughting command menu showing main features.

current position to any designated position on the screen.

(iii) Delete options — e.g. DEL has the effect of deleting lines, circles, arcs, text, fillets, etc., by indicating, with the cursor, the item no longer required.

(iv) Set up commands — TYPE, for example, sets up the type of line required (e.g. chain, dotted, solid).

(v) Feature commands — These commands save time during the construction of a drawing. RDRW, for example, redraws the view on the screen after editing.

(3) Use the feature command ASYM to recall two pre-defined symbols; BOXA3 (used to define the outer edges of the drawing) and BORDA3 (a standard drawing sheet giving machining details and company headings, etc.).

(4) Select GRID and set to 10 mm size. This will allow points and lines to lie on a grid for ease of drawing. Construction grids can be uniquely spaced in two axes and the units of spacing can reflect the *ACTUAL* geometry of the component. Geometry created using grid points stays in those positions, even if the grid is switched off or the spacing altered. The grid points are merely an aid to drawing production and are not plotted on the finished drawing. Line ends can

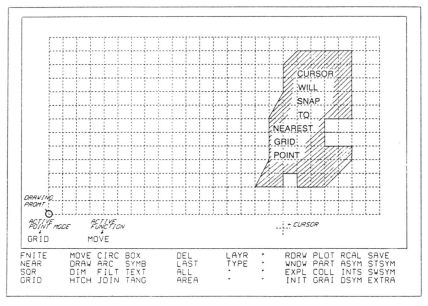

Fig. 4.13. Drawing grid (the cursor will snap to any corner point of grid to aid construction of drawing).

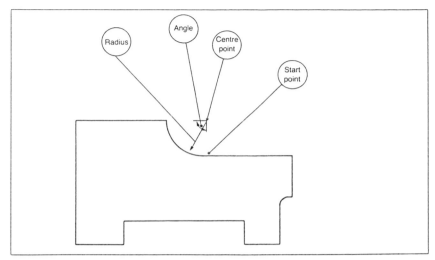

Fig. 4.14. Outer profile of component showing how ARCS are generated.

be positioned (*SNAPPED*) on the nearest grid point.

Figure 4.13 shows the grid on the screen.

(5) Select an appropriate drawing layer for production of the drawing (the layers are colour coded for ease of use). The outline of the component can now be produced on layer 0, by positioning the cursor and using combinations of the commands MOVE, DRAW and ARC.

Arcs for this drawing were produced by defining the centre point, the start point of the arc and radius, and the angle of the arc as shown in Fig. 4.14.

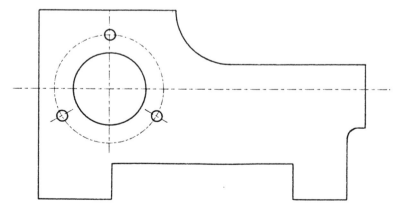

Fig. 4.15. View to show generation of circles and centre lines.

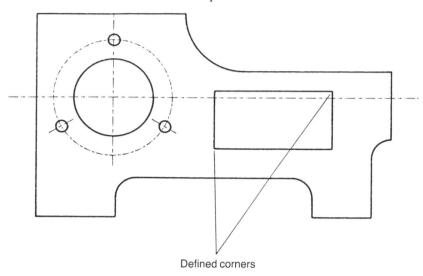

Defined corners

Fig. 4.16. Slot produced by defining two opposed corners of a box.

(6) The three small holes in the component can now be drawn by defining a circle and repeating it three times on a given pitch circle diameter. The large hole is drawn by simply defining its position and radius. A centre line is drawn on a separate layer by defining a circle and selecting the appropriate line type. See Fig. 4.15.

(7) The slot is drawn by selecting the function command, BOX, enabling a box to be drawn by indicating, with the cursor, two diagonally opposed corners. Figure 4.16 shows the slot in position.

(8) The drawing for one view is now complete — all that remains is to supply dimensions and annotations.

    To dimension a side, for example, select an appropriate drawing layer (layer 3 in this case) and select DIM from the menu. By hitting the 'S' key the dimension text size may be re-set to a size required. Provide the computer with *three* points; two points near to the ends of the side to be dimensioned (position of leader lines), and the desired position of the dimension, as shown in Fig. 4.17. The dimension will be calculated and echoed on the screen. All other dimensions are automatically calculated (by the system searching its database) and then drawn on the screen by simply indicating three points for each dimension.

    Text can be entered by selecting the TEXT option. Text can be

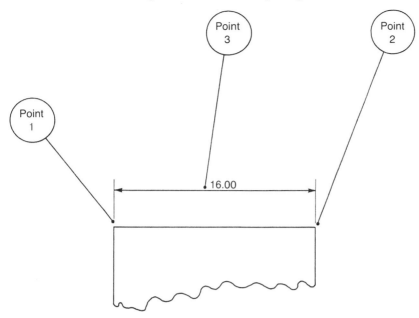

Fig. 4.17. Construction of dimensions by providing three points on the screen.

drawn in any position, at any size or angle. Text can be selected to be drawn left justified, right justified, centred or fitted between two prescribed points. Figure 4.18 shows the drawing fully dimensioned together with text.

(9) To render the drawing more useful, a sectioned view can be easily constructed to provide additional detail. Layer 4 is selected and the INTERSECT mode is used to aid production of a new view by finding the point of intersection between two lines or between a line and a circle. BOX is used to create the drawing outline in the position indicated by INTERSECT, as illustrated in Fig. 4.19.

(10) The view can be sectioned by selecting HATCH and indicating the corners of the boxes to be sectioned. The default hatch angle is 45° as shown in Fig. 4.20.

(11) This view is dimensioned using the techniques described in procedure No. 8.

(12) To complete the working drawing the symbol BOXA3 is swapped for BORDA3 using the command SWSYM. The REDRAW command is initiated and additional text is placed on the drawing. The drawing is then saved away to disk or additionally may be directed

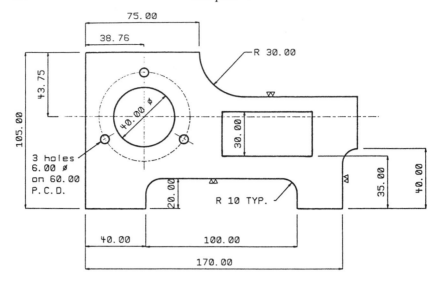

Fig. 4.18. The drawing with dimensions and text.

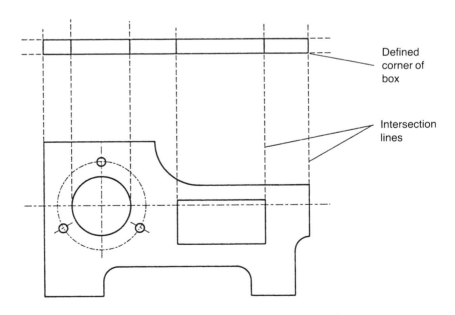

Defined
corner of
box

Intersection
lines

Fig. 4.19. Construction of additonal view using the INTERSECT mode.

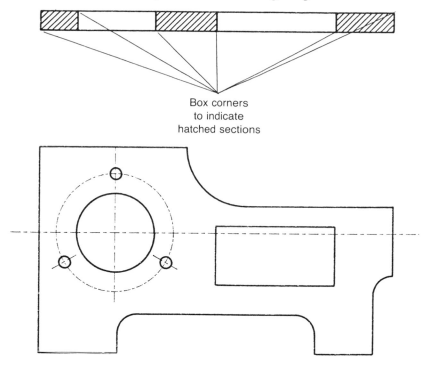

Fig. 4.20. Hatched view generated by indicating corners of BOXes.

to a plotter using the PLOT command. The final plotter produced drawing is given in Fig. 4.21.

The information contained in layer 1, i.e. the drawing outline, can be called together and saved as a separate symbol. This is useful if the 3-D visualization program is required.

(13) Finally, exit from *MAEDOS* program.

4.5 3-D GEOMETRY VISUALIZATION SYSTEMS

These systems have the ability to define three-dimensional geometries in the form of a single model which may be viewed from any given angle or direction. The use of geometry visualization systems provides a three-dimensional image of a component giving a more creative and useful aspect of the 2-D drawing.

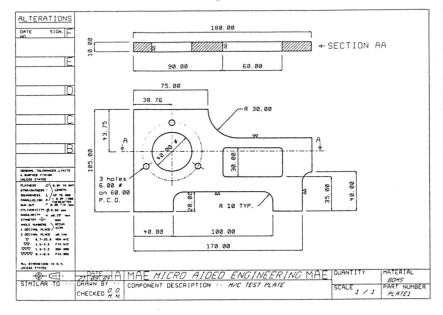

Fig. 4.21. The final working drawing with border symbol (BORDA3).

Typical System

One system which provides 3-D views of 2-D drawings is *MAEVIS* from Micro-Aided Engineering. *MAEVIS* is a three-dimensional visualization system that interfaces with *MAEDOS* to give the user the opportunity to look at perspective views of design elements built up from the *MAEDOS* plan and elevations.

To use *MAEVIS*, the software is loaded and a symbol name is called up from disk. A question and answer session then follows which determines the size of the object to be viewed and the angle and direction from which it is to be viewed. It is usual to enter the location height and extrusion depth of the component, together with the viewing angles and azimuth value. The line type and pen type may also be selected at this stage.

The image, which is essentially a 2-D shape swept into a 3-D volume by parallel motion, can now be viewed on the screen or alternatively can be plotted via *MAEDOS*.

Figure 4.22 shows a 3-D visualization, generated using *MAEVIS*, of the component created by *MAEDOS*.

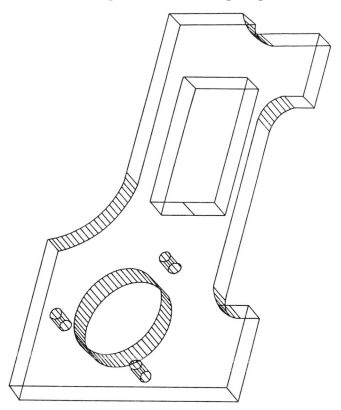

Fig. 4.22. 3-D visualization of 2-D component drawn using *MAEVIS* system.

4.6 ADVANTAGES AND LIMITATIONS OF COMPUTER-AIDED DRAUGHTING

By reducing the demands of the pencil and paper skills of the draughtsman, a CAD system permits greater concentration on the design and layout aspects of the drawing task.

As demonstrated in this chapter, the basic principle of computer-aided draughting is to create a drawing image on a screen, to optimize the design by editing and manipulating, and then output the finalized image on a plotter.

In the course of this process, edited images are stored on disk and this feature in itself provides many advantages.

The main benefits of CAD draughting over manual methods include:

(i) Because the plotter has a better definition than drawing instruments, it is possible to increase the information density of drawings. This includes improved accuracy and presentation (line clarity, lettering, etc.).

(ii) Time can be saved by using previously defined symbols.

(iii) Time can be saved by editing existing drawings rather than producing a new drawing.

(iv) Vast savings can be made on conventional drawing materials and print machines.

(v) More convenient to store drawings on disk than in a bulky drawing cabinet.

(vi) The drawing quality improves the market image of the Company.

(vii) One of the major benefits of draughting systems lies in the ability to link CAD with CAM.

There may be disadvantages associated with computer-aided draughting but they are small compared with the advantages.

Some disadvantages include:

(i) High initial capital outlay.

(ii) Less flexibility in certain drawing environments.

(iii) Equipment may be under utilized, which could reflect the original purchasing requirements.

(iv) Companies are dependent on machinery which can break down.

Chapter 5

Three-Dimensional Modelling

5.1 INTRODUCTION

In many industries there is a growing need for methods of processing information which are related to the *actual shape* of mechanical components. One reason for this stems from the increased use being made of computers in design and manufacturing activities.

The conventional way of conveying shape information, i.e. by engineering drawings, has certain drawbacks in connection with the use of computers. Not only are drawings sometimes unsuitable for direct entry to a computer, but they rely upon human interpretation to recognize three-dimensional shapes from combinations of two-dimensional projections. Draughting is essentially a 2-D activity containing many repetitive tasks and is therefore suitable for improvement using computer.

The engineering designer, on the other hand, faces the more demanding job of creating a three-dimensional component which must perform a specific task. The component will have to meet certain geometric and operational constraints and may also need to be aesthetically styled. Although designers normally record their designs in 2-D (i.e. on a drawing board), they will often be seen producing three-dimensional paper, clay and wood models, and on occasions will have to produce expensive prototypes. Traditionally, there has been a tendency to produce items of a simple shape, but increasing market pressure is forcing companies to produce components which have more complex geometry.

It is not until the first prototype becomes available that the designer is able to get full three-dimensional appreciation of the component design. There can be problems with interference on assembly or operation with prototypes. Calculation errors can also cause expensive re-design. The more complex the design the less likely it is that these faults can be remedied by re-design and instead certain compromises are accepted.

The use of a 3-D computer model for the initial design would have identified most of the problems as they occurred and long before the production of expensive prototypes. The use of a 3-D modeller results in the following benefits: reduced product cost, improved product quality and reliability, and a reduction in lead-times giving increased overall profitability.

This chapter explains the main types of 3-D modeller, and describes the fundamental procedures for creating solid models. A case study is included to show how a simple component can be built up from primitive shapes. The ways in which 3-D modellers are used to simulate dynamic mechanisms is discussed and typical application areas of modelling are outlined. Finally, the major benefits and limitations of modellers are given.

5.2　3-D MODELLERS

Research into 3-D modellers began in the mid 1960s when universities and companies with large computing resources investigated *wireframe* or stick figure models.

Surface modelling was investigated by aerospace and automotive industries as this type of model is particularly useful in styling. *Solids modelling* research continued through the 1960s and 1970s until the first commercially available solids modellers reached the market place towards the end of the 1970s. Today, many CAD/CAM suppliers offer modelling capabilities and growing attention is now being paid to the use of solids modellers, in particular, as a significant design and manufacturing tool. A 3-D model can be built up by the user having an imaginary cube which exists inside the computer memory, the size of the cube dictating the limits of the model. The cube has a *global* co-ordinate system as shown in Fig. 5.1.

The use of global co-ordinates, whilst ideal for the majority of construction work in modelling, can be restrictive if a user wishes to work only on a small part of the master model. For this reason modellers have 'local co-ordinate' facilities which allow the user to work relative to a re-defined local origin or plane.

The precision of the model is limited, essentially, by the computational accuracy of the computer. It is usual for the system to operate in double precision arithmetic since round off errors induced by geometric transformations could give rise to unacceptable accuracy in single precision arithmetic.

Modern 32-bit computers can work in double precision floating point notation which gives extremely high accuracy.

Models can be built in real units, i.e. mm, inches, km, the representation of the model being scaled for viewing purposes.

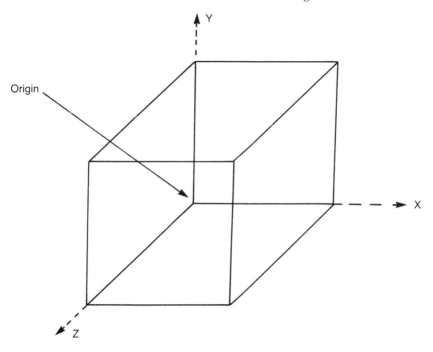

Fig. 5.1. 3-D modelling cube with origin and global co-ordinates.

Types of Modeller

Three-dimensional objects can be modelled using one of three methods: Wireframe modelling, surface modelling or solids modelling.

Wireframe Models

A wireframe model of an object is made up of a set of three-dimensional co-ordinates, which define the end points of lines in space. Information about the type of line is also held (i.e. the curvature of the line). This type of model, because it is described by its edges and vertices, can only provide *partial* information. Models have the appearance of a frame of wires as shown in Fig. 5.2.

The wireframe model is built up to represent the edges of an object but additional information cannot be added to identify which lines actually represent the edge of a surface. The model cannot be used to perceive any information about the volume of the object or indeed

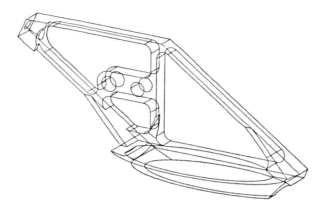

Fig. 5.2. Wireframe display of 3-D object model (reproduced by kind permission of Shape Data Ltd).

which lines represent the inside or the outside of the component.

With many wireframe models hidden lines cannot be removed and so the displayed wireframe image can be *visually ambiguous*. It is the user's responsibility, therefore, to ensure that no edge lines are missing.

Despite the fact that the majority of wireframe models do not provide volume or mass property details, they are still widely used in industry as this type of model is perfectly adequate for representing views which are difficult to interpret from a two-dimensional drawing. Additionally, the computer memory requirement for a wireframe model is not a great deal more than that of a 2-D drawing.

Surface Models

Surface models hold the description of an object in terms of points, edges and faces between the edges. Surface modelling systems are able to calculate surface intersections and surface areas and some systems are ·capable of producing shaded images and removing hidden lines automatically.

Essentially, the models produced by these systems are in the form of a *mesh*, constructed from a set of measured or calculated co-ordinates. These co-ordinate points which describe the form of the surface of the object are input to the system which uses this geometry data to create a meshed surface. The fineness of the mesh can be specified by the user.

Surface models are ideally suited to applications where the model of a complex surface is required, such as aerospace and car body styling. Modern surface modellers are capable of producing vivid colour shading which gives a unique level of realism to the object displayed.

Many systems provide for the creation of a variety of surface types by entering single commands at the workstation. Some surface modellers offer a tube or pipe surface component which takes a two-dimensional profile and effectively sweeps it perpendicular to a straight or curved line to rapidly produce a complex surface.

Figure 5.3 shows a component produced using a surface modeller.

From the geometry of a surface model, numerical control tool path data can be generated. All machining operations, including multi-axis profiling can be catered for, allowing complex surfaces to be machined.

As the model is made up of a mesh, the surface is, in fact, a collection of small faceted surfaces. Therefore, for NC machining, the

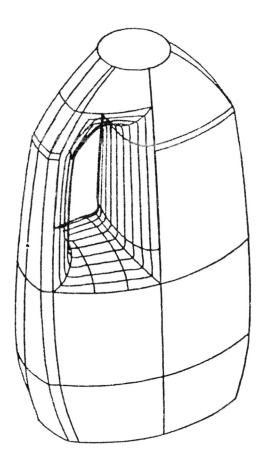

Fig. 5.3. Output from DUCT surface modeller (courtesy Delta CAE Ltd).

mesh size has to be sufficiently fine to enable smooth continuous surfaces to be machined. In this way accurate moulds and patterns can be
produced.

Surface modelling can be used to optimize component weight for
both geometry and stress. There are limitations with some surface
modelling in that there is no information to describe which parts of the
model are solid or indeed which is the inside or outside of the object.
The creation of sectional views can also be a time-consuming task.

Solid Models

A solids modeller is probably the most powerful of the three-dimensional modelling techniques as it provides the user with complete information about the outline, surface, volume and mass properties of the
model. A solids modeller holds a complete description of an object, in
terms of the space which it occupies. The description can indicate
whether any point is inside our outside the object or whether it lies on
the surface of an object. Solids modellers build up components from
three-dimensional shapes that can be described mathematically. As
these systems produce complete and unambiguous geometric models
of objects, they require a minimum of human explanation or additional
information: they may provide a foundation, therefore, for truly automated engineering. Figure 5.4 shows a component generated using a
3-D solids modeller.

Two fundamental types of solids description method have evolved.
The first is the *constructive solid geometry* (known as CSG) data structure which models the simple elements needed to build up the component, together with the operations to combine them. Such a data structure is very compact but does not represent the final edges and faces
which are produced after combining the simple elements. The whole
model has to be re-evaluated after each design change, and interaction
with these systems can be difficult.

The second is the *boundary representation* method (B-rep) which
uses a boundary file data structure. With this approach every face, edge
and vertex of the final object is represented. The data structure is
designed to define how these details are connected together along with
the geometry of each detail. The B-rep method provides fast image
generation, direct user interaction, and a variety of modelling operations.

The engineering designer can gain enormous benefits from a solids
modeller at the conceptual stage in design, as information concerning

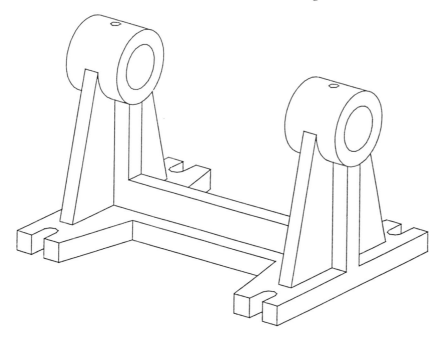

Fig. 5.4. Solids model (courtesy Leeds University Geometric Modelling Project).

mass, centre of gravity, and moments and products of inertia are available in a fraction of the time it would take to calculate manually. For detail design the solids modeller can be used to check for interference between two component parts. The system can be made to display the volume of overlap between two parts following assembly. Since the full boundary information about the model is available, it is possible to process this data to produce NC machining tapes or a finite element mesh description. Solids modelling systems can produce very impressive visual displays of objects, using perspective, colour shading and highlighting to create images of photographic quality. These images can be used in advertising leaflets or technical handbooks.

The computer storage required to hold the geometric definition of an object can be quite large (typically 400 K bytes).

5.3   CREATING 3-D SOLID MODELS

This section describes the way in which a constructive solid geometry (CSG) model is built up and manipulated.

CSG modellers describe objects in terms of solid primitive shapes which may be combined in a variety of ways. Primitive shapes usually include BLOCKS, CONES, CYLINDERS, SPHERES and TORI as shown in Fig. 5.5.

The model is built up by the user defining the type of primitive required, the key dimensions, its position relative to the origin (or relative to its current position) and its relationship (e.g. how it is to be combined) with primitives already created.

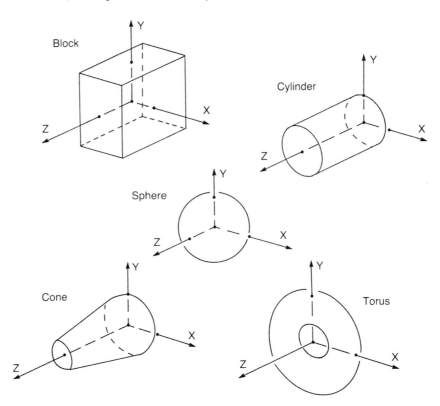

Fig. 5.5. Modelling primitives.

Operators

The relationship between primitives are defined using the BOOLEAN operators UNION, DIFFERENCE and INTERSECTION.

These operators are derived from basic set-theory and in the context of geometric modelling this means that solids can effectively be added, subtracted and shared. While these operators are not in themselves

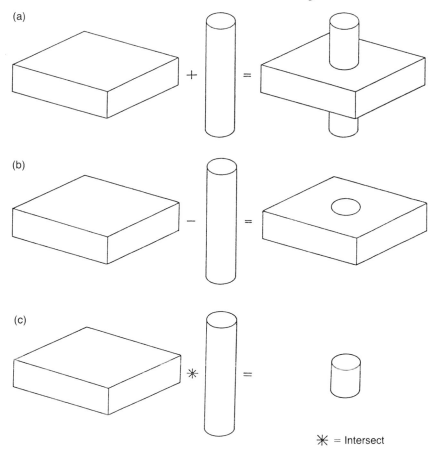

Fig. 5.6. Effect of Boolean operations on primitives (a) UNION
(b) DIFFERENCE
(c) INTERSECTION

sufficient or convenient for all mechanical design, they constitute a powerful tool for shape manipulation.

The effect of a Boolean operation can be most easily visualized in terms of those points which lie within the solid part of the model.

The UNION operator combines two bodies into a single new body as shown in Fig. 5.6a. This combination can now be treated as a single homogeneous body. Any point in either of the original bodies becomes a point in the new body.

The DIFFERENCE operator may be used if the removal of material from a body is required as illustrated in Fig. 5.6b. A DIFFERENCE operation applied to two bodies removes from the first those parts common with the second.

The INTERSECTION operator has the effect of calculating the common volume shared by two bodies. The result of an INTERSECT operation contains only those points that lay within both original bodies, as shown in Fig. 5.6c.

Parts can be built up using combinations of these operators and the parts are given names as they are built. This is very useful if the whole assembly is to be modified at a later date. Each body can be individually moved (in *x, y* and *z*), rotated, or its dimensions changed without the need to re-build the entire model.

Spinning a Profile

To create a model which is symmetrical about an axis, a two-dimensional profile can be built up and then spun (usually by 360°) about an axis, as shown in Fig. 5.7. This technique can be used to save time in modelling but will consume more storage space than if the object were created using a series of primitives.

Either an open or closed profile may be spun around an axis to form a solid of revolution:

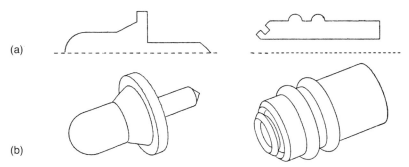

(a)

(b)

Fig. 5.7. (a) Half section of components.
           (b) Spun profiles.
(Reproduced by kind permission of Shape Data Ltd.)

Assemblies

Assemblies are collections of components which are fitted together without the components becoming a single object. Models can be built up by assembling a series of primitives or objects. The individual components are not combined as with the UNION operator but instead

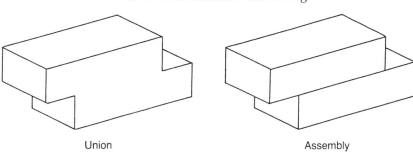

Union   Assembly

Fig. 5.8. Distinction between UNION and ASSEMBLY.

retain their own identity.

Figure 5.8 illustrates the distinction between UNION and ASSEMBLY.

Viewing

Models can be viewed from any position in three-dimensional space. In this way the user can obtain a selection of views in order to gain a full appreciation of the model. Most solid modelling systems allow 3-view orthographic projections of the model. One object may be rotated inside another object, for example, to visualize the movement of a butterfly flap within the valve housing.

Sectioning

The production of a sectioned view of a solid model is quite straight-forward. All that is required is to cut the model with a sectioning plane by differencing it with a simple primitive, as shown in Fig. 5.9. The section revealed can either be cross-hatched or left blank.

There are further features for viewing 3-D models including hidden line removal, colouring and shading. Assemblies may also be shown as exploded views for inclusion in technical and sales leaflets (see Fig. 5.10).

5.4   MODELLING CASE STUDY

The object of this case study is to show how a reasonably simple mechanical engineering component can be modelled using one type of geometric modeller.

(a) Cylinder   (b) Block

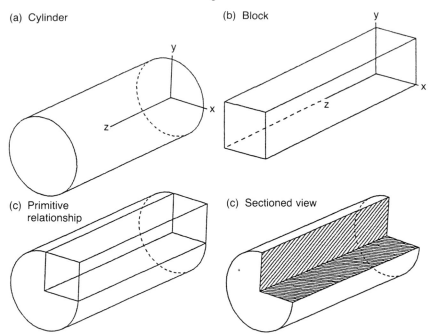

(c) Primitive   (c) Sectioned view
    relationship

Fig. 5.9. Cylinder differenced with a block to produce sectioned object.

The software to be described is NONAME, which is a constructive solid geometry 3-D modelling system developed in the Mechanical Engineering Department at the University of Leeds.

Objects are defined by series of typed statements, entered at the workstation, describing how each can be built up by applying Boolean operations to primitive shapes. For easy readability, primitives, parameters, and combinations of primitives can be given suitable descriptive names.

The primitives are:

BLOCK(x length,y length,z length)	Origin at centre
CYL(height,radius)	Origin at centre of bottom face, lying along z axis
CONE(height,top radius,bottom radius)	Origin at centre of bottom face, lying along z axis
SPHERE(radius)	Origin at centre
TORUS(minor radius,major radius)	Origin at centre, lying in xy plane

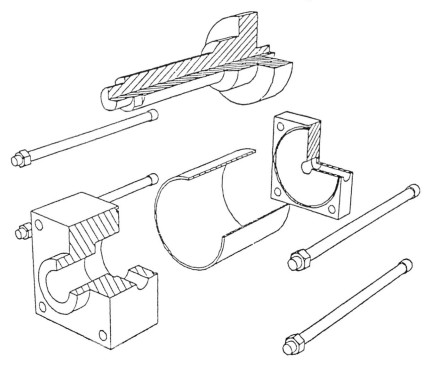

Fig. 5.10. Exploded view of a solids model (courtesy Leeds University Geometric Modell-ing Project).

In order to assign names to indicate primitives the symbol <- is typed.

Examples of complete primitive statements are:

    BL<-BLOCK(30,30,40)
    TAPER_1<-CONE(17,4.5,15)
    BEARING_TOP<-SPHERE(130.6)

Primitives can be positioned relative to the x, y, z origin using state-ments of the form:

    TAPER_2<-CONE(17,4.5,15) AT (x posn,y posn,z posn)

and operators are used in the following form:

    BIG_BLOCK<-BL1+BL2   for UNION
    SMALL_BLOCK<-BL1-BL2   for DIFFERENCE
or, INT_BLOCK<-BL1*BL2   for INTERSECTION

The system comprises a number of modules to create models, produce drawing layouts, evaluate mass properties, and derive models from linear and rotational sweeps.

NONAME software is designed to run on 32-bit mini and mainframe computers.

The component to be modelled is the ARM shown in Fig. 5.11.

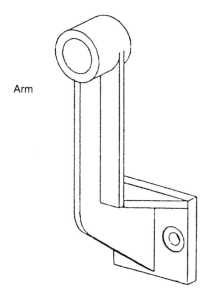

Arm

Fig. 5.11. ARM (courtesy Leeds University Geometric Modelling Project).

Modelling Procedure

1. Define three primitives, a block named A_BLOCK and two cylinders named A_CYL1 and A_CYL2,
   where: A_BLOCK<-BLOCK(60,180,8)AT(0,−90,16)
          A_CYL1<-CYL(57,30)
          A_CYL2<-CYL(57,20)

2. Perform Boolean functions on the primitives to produce a definition of an object named A_BOSS, as shown in Fig. 5.12, where:
   A_BOSS<-A_BLOCK+A_CYL1−A_CYL2

3. Save definition to disk under filename BOSSY.DEF

4. Define two cylinders IN_CYL and OUT_CYL,
   where: IN_CYL<-CYL(14,7.5)

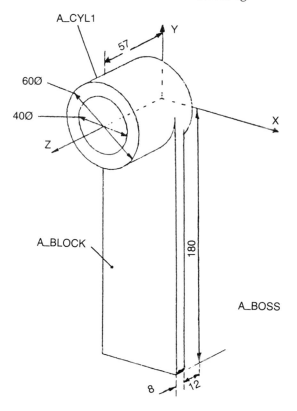

Fig. 5.12. A_BOSS made from a cylinder and a block (courtesy Leeds University Geometric Modelling Project).

OUT_CYL<-CYL(2,15)AT(0,0,14)

5. Use these cylinders to produce an object called A_HOLE, as shown in Fig. 5.13, where: A_HOLE<-IN_CYL+OUT_CYL

6. Replace the single hole by two holes A_HOLE_1 and A_HOLE_2. These two holes are the same as the original but are moved to new positions, as shown in Fig. 5.14. Call this object TWO_HOLES, where: A_HOLE_1<-MOVE(A_HOLE)BY(MOVEX=-30)
      A_HOLE_2<-MOVE(A_HOLE)BY(MOVEX=30)
   and TWO_HOLES<-A_HOLE_1+A_HOLE_2

7. Define a block named BIG_BLOCK and produce an object named A_PLATE, as in Fig. 5.15, by differencing TWO_HOLES from BIG_BLOCK,
   where: BIG_BLOCK<-BLOCK(100,100,16)AT(0,0,8)

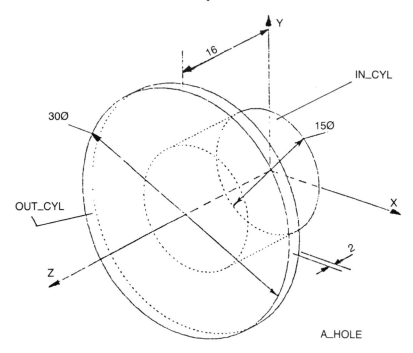

Fig. 5.13. A_HOLE made from two cylinders (courtesy Leeds University Geometric Modelling Project).

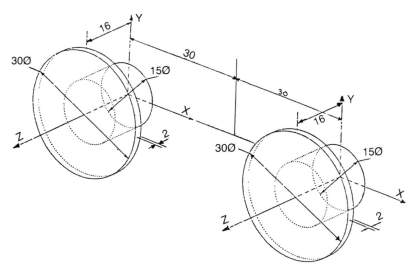

Fig. 5.14. TWO_HOLES produced by moving the object A_HOLE (courtesy Leeds University Geometric Modelling Project).

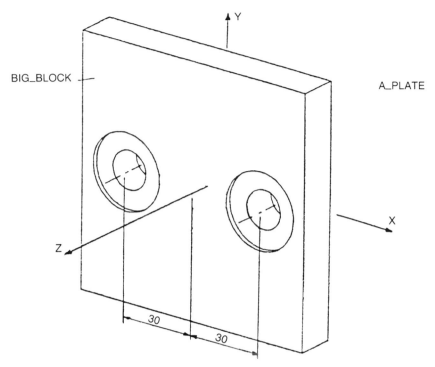

Fig. 5.15. A_PLATE produced by differencing a block with TWO_HOLES (courtesy Leeds University Geometric Modelling Project).

and   A_PLATE<-BIG_BLOCK-TWO_HOLES

8. Save definition under filename PLATY.DEF

9. Create an object named FLNG_BL1, as in Fig. 5.16a, where:
   FLNG_BL1<-BLOCK(100,8,100)
                     AT(MOVEY=4,MOVEZ=50,ROTX=-15)

10. Create two blocks FLNG_2_B2 and FLNG_2_B3, and difference them from a block named FLNG_2_B1 to produce an object named FLGN_BL2, as shown in Fig. 5.16b, where:
    FLNG_2_B2<-BLOCK(30,40,100)AT(65,20,50,ROTY=-12.5)
    FLNG_2_B3<-BLOCK(30,40,100)AT(-65,20,50,ROTY=12.5)
    FLNG_2_B1<-BLOCK(100,40,90)AT(0,20,45) , and
    FLNG_BL2<-FLNG 2_B1-FLNG_2_B2-FLNG_2_B3

11. Create A_FLANGE, as shown in Fig. 5.16c, by intersecting FLNG_BL1 and FLNG_BL2,
    i.e. A_FLANGE<-FLNG_BL1*FLNG_BL2

12. Save definition under filename FLANGY.DEF

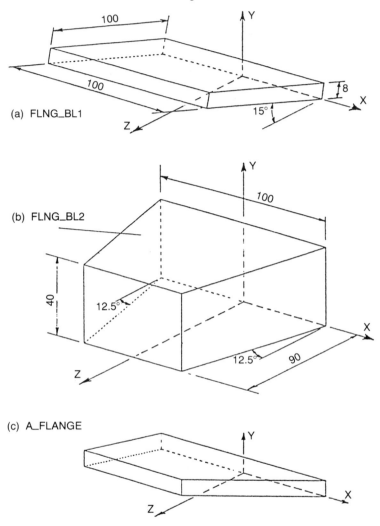

Fig. 5.16. Components of a A_FLANGE (courtesy Leeds University Geometric Modelling Project).

13. Create two blocks PART_WEB_1 and PART_WEB_2, as shown in Figs 5.17a and 5.17b. Intersect these blocks to produce INT_WEB, as illustrated in Fig. 5.17c, where:

    PART_WEB_1<−BLOCK(8,125,196)AT(0,62.5,0)
    PART_WEB_2<−BLOCK(10,74,150)AT(0,0,−75,ROTX=−35)
    and   INT_WEB<−PART_WEB_1*PART_WEB_2

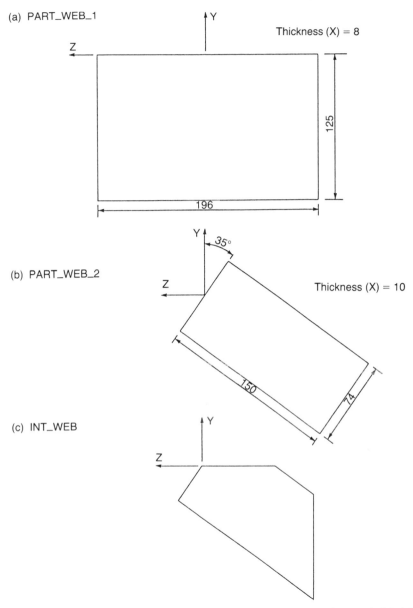

(a) PART_WEB_1

Y

Thickness (X) = 8

Z

125

196

(b) PART_WEB_2

Y

35°

Z

Thickness (X) = 10

150

74

(c) INT_WEB

Y

Z

Fig. 5.17. Components of INT_WEB (courtesy Leeds University Geometric Modelling Project).

14. Create INT_2_WEB by adding a block 2B_WEB and a cylinder 2C_WEB to INT_WEB, see Fig. 5.18a, where:

2B_WEB<-BLOCK(8,150,74)AT(0,75,0)
2C_WEB<-CYL(8,37)AT(MOVEZ=-4,ROTY=-90) , and
INT_2_WEB<-INT_WEB+2B_WEB-2C_WEB

15. Create A_WEB by differencing two blocks, PART_WEB_3 and
   PART_WEB_4, with INT_2_WEB, as shown in Fig. 5.18b.

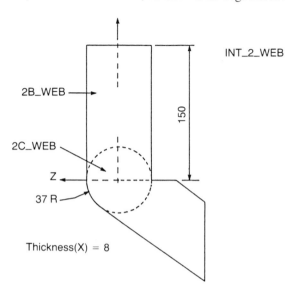

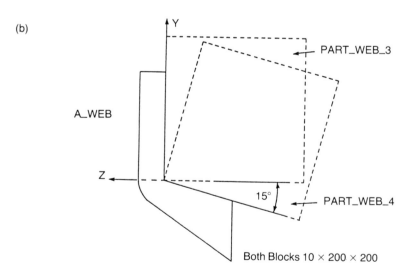

Fig. 5.18. Construction of A_WEB (courtesy Leeds University Geometric Modelling Project).

16. Save the definition under filename WEBBY.DEF

17. Recall files BOSSY.DEF, FLANGY.DEF, and PLATY.DEF from disk. Create new components A_PLATE_2, A_WEB_2 and A_FLANGE_2 by moving A_PLATE, A_WEB and A_FLANGE respectively to new positions relative to the global origin. In Fig. 5.19 the initial position of each component is shown in dashed lines and the new position is shown in solid lines.

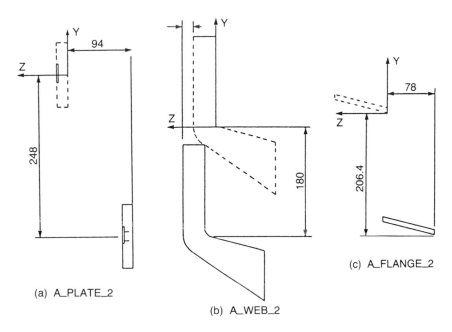

(a) A_PLATE_2

(b) A_WEB_2

(c) A_FLANGE_2

Fig. 5.19. New objects created by moving A_PLATE, A_WEB, and A_FLANGE (courtesy Leeds University Geometric Modelling Project).

18. The component parts of the arm have been created and it is now possible to assemble them together to produce the desired model. Assemblies are created using the symbol <<– therefore,

ARM<<–A_BOSS & A_PLATE_2

& A_WEB_2 & A_FLANGE_2

will produce the assembly shown in Fig. 5.11.

19. To give a clearer interpretation of how the component parts of the arm fit together it is useful to have a sectional view of the arm. Sections are created using the symbol <<<–.

To view a half section of the arm along the *y* axis, the arm model can be sectioned with a suitable sized block, i.e.

ARM_SEC<<<–SECT ARM WITH
$$\text{BLOCK}(60,400,250)\text{AT}(30,-150,0)$$

to produce the view given in Fig. 5.20.

20. Exit from the NONAME program

The case study presented here has utilized only a tiny proportion of the many features and facilities available in NONAME. To demonstrate the more advanced features of this software would require a book in its own right.

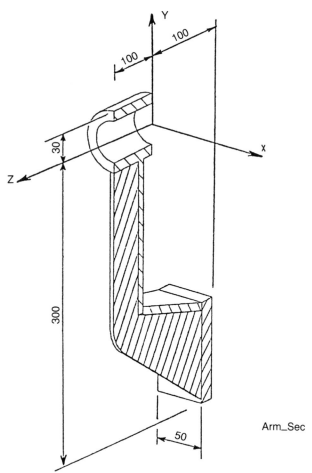

Fig. 5.20. View of arm revealed after sectioning with a block (courtesy Leeds University Geometric Modelling Project).

5.5 PARAMETRIC DESIGN

Engineering companies usually have expertise in one particular field of engineering. As a result they manufacture components that can be varied to suit individual customer requirements. Although the components are, in the main, comprised of the same elements, the dimensions and shape may differ from one application to another.

There are a variety of routines within modeller software which cater for parametric design of components together with design and kinematic analysis of mechanism and linkages.

Component Design

The procedure is to design a component to *standard* dimensions and then assign a unique code to every feature or dimension that is likely to vary. This procedure creates a parametric design master, and a user only has to enter a set of values for the coded features and dimensions in order to produce a given component. A parametric design master is illustrated in Fig. 5.21. This figure shows a lever, where all the variable features are coded using descriptions of the features. Figure 5.22 depicts a dimensioned version of the master shown in Fig. 5.21.

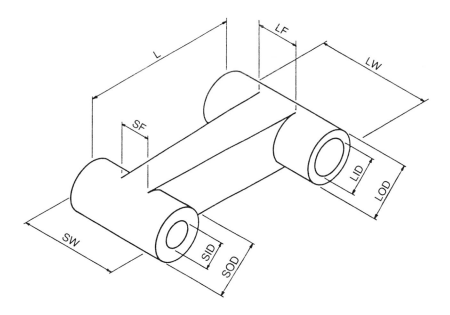

Fig. 5.21. Parametric master.

```
PARAMETER LIST
==============

enter parameters

L=?    12U
SV=?    30
SF=?    1U
SID=?  12.5
SOD=?  15.75
LW=?   1U0
LF=?    5U
LID=?   8.5
LOD=?   4U
```

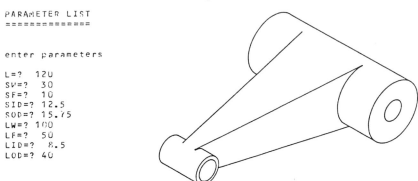

Fig. 5.22. View of component following entry of parameters.

Parametric design techniques can usefully be applied to other areas of engineering design. With a mechanism or linked structure, for example, it is often essential to have information at various stages of motion. It is a lengthy exercise to design linkages and mechanisms manually as a sequence of layout drawings showing various positions of the component parts is required.

Linkage Analysis

The design and analysis of linkages can be carried out by first modelling the complete linkage in a given position and then entering parameters for angles of movement of the arms about pivot points and then automatically re-drawing the linkage in its new position. This procedure can be programmed in a loop so that the complete cycle of the linkage motion can be shown.

Figure 5.23 shows a simple piston linkage in a set position and Fig. 5.24 illustrates how the complete motion of the piston and linkage can be plotted using parametric design techniques.

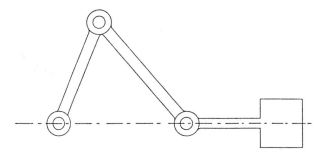

Fig. 5.23. Piston linkage in set position.

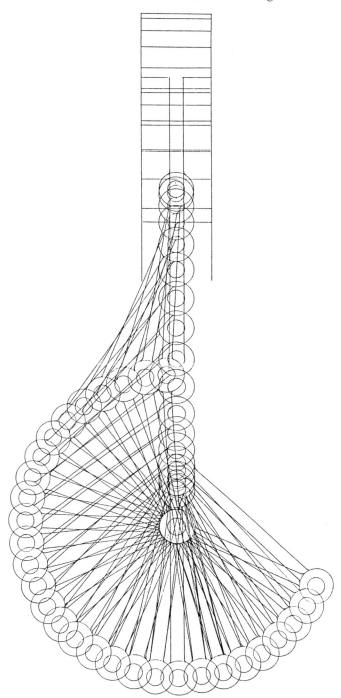

Fig. 5.24. Plotted motion of linkage using Parametric Design techniques (reproduced by kind permission of Cambridge Interactive Systems (Products) Limited).

Mechanisms

In the design of mechanisms it is necessary to be able to define the path of all component parts. Using parametric design it is possible to rotate one part of the mechanism and detect when it comes into contact with another part, for example. It is also possible to detect when a clash of parts might occur. Figure 5.25 shows a Geneva mechanism in various stages of operation. The input pinion can be rotated about its pivot and the resulting action of the follower disk can be analyzed. Parametric design is a very useful tool when designing housings for mechanisms and linkages.

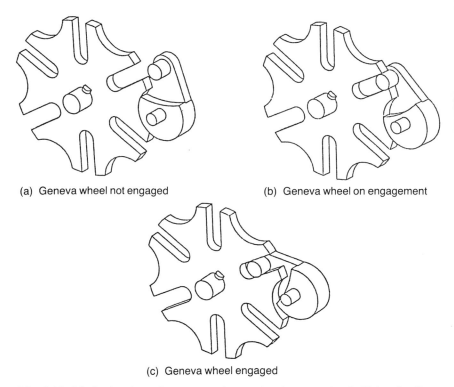

(a) Geneva wheel not engaged   (b) Geneva wheel on engagement

(c) Geneva wheel engaged

Fig. 5.25. Mechanism in various stages of operation (courtesy Leeds University Geometric Modelling Project).

5.6   APPLICATIONS OF 3-D MODELLERS

Three-dimensional modellers can be used for a variety of engineering applications. The modeller can be used on its own for design and visualization of objects, or can be interfaced with other CAD/CAM systems so that the geometric definition of the model can be down-loaded for use in draughting, finite element analysis, manufacturing and robotics.   Typical uses of 3-D modellers include:

(i) *Design*

The engineering designer is able to visualize the component under development. It is easy to produce sectioned views in order to check for wall thickness or hole intersections. Errors of interference on assembly are readily identified. The model can be used to calculate *engineering mass properties*, such as volume, weight, centroids, and products and moments of inertia. Parametric design can help the understanding of the behaviour of mechanisms and linkages, and effective use of the modeller will make it easy to accommodate design changes and will reduce the need for expensive prototypes.

(ii) *Visualization*

A modeller can be a valuable asset to the marketing department of a company. An assembly of components may be modelled and the image exploded to provide clarity of information. Exploded views give an improved understanding of the design of a product.

Information is conveyed more readily by the use of 3-D views than by the use of 2-D drawings alone.

Where colour is a relevant factor in design, the geometric modeller allows many colour alternatives so that parts of the model can be evaluated easily. Models can be shaded, illuminated and viewed from any angle.

Images, like the one shown in Fig. 5.26, can be reproduced for inclusion in technical handbooks or sales literature.

To exploit fully the benefits of 3-D modellers, it is possible to interface the modeller with other CAD/CAM activities. This involves extracting the geometry definition data of the model, re-formatting the data and then downloading the information to a desired CAD/CAM application.

(iii) *2-D Draughting*

It is possible to produce engineering drawings from a 3-D modelling system. Data can be downloaded to a draughting system so that orthographic and sectioned views can be generated and used by companies for conventional engineering manufacture.

There is usually a two-way interaction which enables 2-D draughting information to be input to the 3-D modeller.

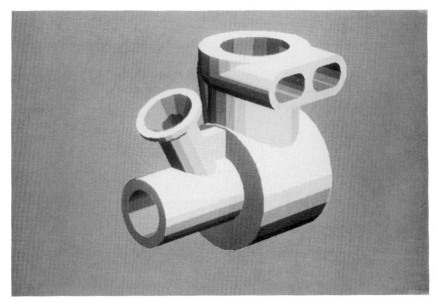

Fig. 5.26. Monochrome photograph of 3-D image which has been colour shaded to give maximum visual impact (courtesy Radan Computational Limited).

(iv) *Finite Element Analysis*

The geometric modeller holds all the geometric information necessary for automatic generation of a finite element mesh. The definition data is downloaded from the modeller system to a finite element pre-processor. There is a great deal of research underway to fully automate the generation of FE meshes for analysis.

Figure 5.27 shows a finite element mesh of a 3-D solid model.

(v) *Manufacturing*

Numerically controlled machining is closely allied to 3-D modelling because use can be made of the component geometry data generated by the system. Part-program information can be derived from surface and solid models and used to generate NC toolpaths. The system can be given detail about the types of tool to use, the direction of cut and the tool path details. This additional information is passed via a post-processor to the particular machine tool controller.

Figure 5.28 shows a solid model representation of a part, together with the machined component.

(a)

Solid model

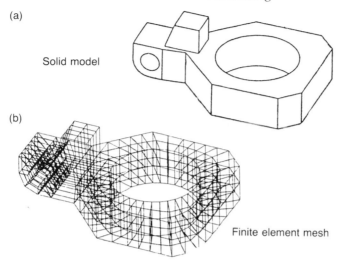

(b)

Finite element mesh

Fig. 5.27. (b) shows the finite element mesh of the model in (a) (courtesy Shape Data Limited).

(a)

Solid model

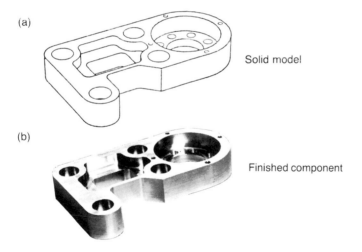

(b)

Finished component

Fig. 5.28. (b) shows the N.C. machined version of the solid model in (a) (courtesy Shape Data Limited).

(vi) *Robotics*

As the shape of an object can be defined in *three* dimensions, this definition can be fed into the robot system. The robot can be programmed to handle a variety of components even before they are actually manufactured. Robot actions can also be simulated using a 3-D modeller, so that two or more robots working in a cell can be checked for interference and possible clashing.

The automatic inspection of parts can also be assisted by transferring geometric definition data to the inspection cell.

5.7   THE BENEFITS AND LIMITATIONS OF 3-D MODELLING

The benefits associated with 3-D modelling are manifold.

Geometric modellers are the focal point of the integration of CAD/CAM systems, enabling engineers and designers to interactively create, modify, test and process design and manufacturing information quickly and efficiently.

More specific benefits include:

(i)   Relatively fast creation of models and quick analysis of design (interference checking, mass properties, etc.).

(ii) Model shop time is significantly reduced or eliminated. Presentations of proposals to management and customers may be made effectively by using a hard copy of the 3-D image.

(iii) Improved visualization of objects for catalogue illustrations are an effective sales or promotional aid. If wireframe models are used for this purpose it is desirable if hidden line removal or shading facilities are available.

(iv) The geometric data definition of components can be used as a parts database for company-wide information. The *potential benefits* to be gained from integrating CAD with CAM via the modelling system cannot be overstated.

(v)   All the design changes, refinements and component analyses are carried out on software and not on hardware prototypes. This results in dramatic savings in time and cost to the design office.

(vi) The graphical prove-out of NC tool paths also saves time and eliminates the possibility of tool/machine clashes.

There are limitations associated with the use of 3-D geometric modellers. The major limitations are:

(i)   3-D modelling software, and indeed the hardware on which it runs, can be expensive items for an engineering company and although the benefits to be gained from such a system are great, the onus is on the individual company to justify the acquisition of 3-D modelling facilities.

(ii) The speed at which the model is created on the screen slows as the complexity of the model grows. In particular, if a CSG solids modeller is used to model an object, the system has to re-calculate the whole model following each design change. For very complicated objects the automatic re-drawing can be time consuming.

(iii) The computation of 3-D models places a heavy burden on the processor and requires a large amount of storage space. 3-D modelling systems, therefore, are only really suitable for use on mini or mainframe computers with 32-bit processors and magnetic disk storage capabilities.

Chapter 6

The Use of Finite Elements as a CAD Tool

6.1 INTRODUCTION

Accurate structural analysis is a vital part of the engineering design process in many world-wide engineering industries, including civil, mechanical, aerospace, automotive, petrochemical and offshore structural design. Engineering designers are often faced with practical problems whose solution by the conventional 'strength of materials approach' is often difficult or impossible as the components under analysis have inherent complex forms.

Faced with this problem, and with industry imposing greater demands on reducing development time and cost, numerical calculation as a design aid has, over the years, been assuming increasing importance. To exploit the benefits of such calculations, realistic and often complex models must be employed to achieve reliable results.

With the arrival of high-speed versatile digital computers, it has become possible to find accurate solutions to a wide range of previously insoluble problems by applying computing power to mathematical techniques.

Many techniques have been tried but none has been more widely used and intensively developed than the *finite element method*. Finite element analysis initially found wide acceptance in the aerospace and defence industries. More recently, however, with trends towards more powerful, less expensive computers and the removal of the mystique which once surrounded the method, it has become available to a large number of small and medium sized companies.

This chapter outlines the finite element method and shows how this technique can be a valuable tool in computer-aided design.

A summary of the underlying principles of finite element analysis is given and two case studies are used to describe the application of the method to the solution of typical engineering problems. Finite element software is discussed and, finally, the benefits and limitations of this type of analysis are outlined.

The Finite Element Method (FEM) has developed simultaneously with the rapidly increasing use of computers in industry, and although the method was originally developed for structural analysis and design, the theory on which it is based now means that the analysis can be used for the solution of problems in engineering disciplines such as biomedical and nuclear.

The *principles* on which the finite element method is based have been understood for over 150 years. The first recorded analysis of what is known as the displacement method is attributable to *Navier*, who in 1826 described the analysis of a simple framework. Progress in the development of theory and the analytical techniques, which eventually led to FE analysis, was particularly slow up to the 1920s. This was due, in part, to the practical limitations imposed on the solution of algebraic equations with a large number of unknowns. During the 1940s however, with military establishments showing considerable interest in the efficient design of aircraft structures, attempts were made to develop methods for analyzing continua. The ideas during this period represented the forerunners of the matrix structural analysis concepts which are adopted for finite element analysis today. It was not until the advent of the digital computer that the solution of large numbers of simultaneous equations became an economic and viable proposition and it was soon realized that structural analysis was one area in which this capability could be exploited.

The technology of the finite element method has advanced quite considerably during the last ten years and the technique is now widely regarded as being one of the most powerful methods of engineering analysis and design.

6.2 WHAT ARE FINITE ELEMENTS?

The finite element method analyzes components by regarding the structure as an assemblage of *discrete* elements of simple shape. These *elements* are considered to be joined together at convenient points on their boundaries known as *nodes*. If the structure has *n* dimensions in space, it can be subdivided into an equivalent system of *n*-dimensional finite elements. The representation of a structure in this way is called *discretization*. One-dimensional bodies may be sub-divided into finite elements by lines between nodes as shown in Fig. 6.1a, whereas polygons and polyhedra may be used for the sub-division of two- and three-dimensional bodies as shown in Figs 6.1b and 6.1c respectively.

In two- and three-dimensional problems it is possible, and often convenient, to represent the body by a combination of triangular, quadrilateral, tetrahedral and hexahedral elements.

Chapter 6

Fig. 6.1a. A one-dimensional body represented by four linear finite elements.

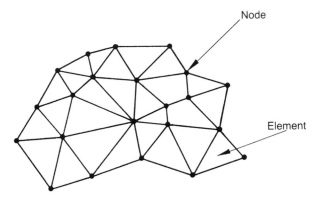

Fig. 6.1b. A two-dimensional body represented by a system of triangular plane elements.

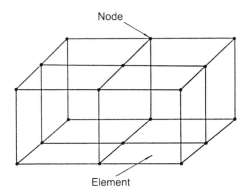

Fig. 6.1c. A three-dimensional solid body represented by four box shaped elements.

    The manner in which the finite elements are attached to their respective nodes can be understood by referring to Fig. 6.2a, which shows three planar finite elements, two triangular and one rectangular. The three elements are separate and not attached in any way to each other. The nodes are used to attach adjacent elements together in order to build up the structure, as shown in Fig. 6.2b. In these simple elements the nodes are always situated at the corners of each element, although more sophisticated elements allow nodes in other positions along their

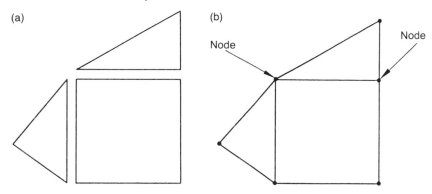

Fig. 6.2a. Three separate plane finite elements.

Fig. 6.2b. Finite elements connected at their boundaries by nodes.

boundaries. If the nodes were removed the elements would separate and there would be no continuity between adjacent elements. Numerous elements can be put together in a variety of ways and they can be arranged to simulate exceeedingly complex shapes. A collection of elements used to represent a structure or component is termed a *mesh*. Figures 6.3a and 6.3b are just two examples of the types of structure which can be analyzed using the finite element method.

One of the major uses of the finite element method is to calculate *displacements* at the nodes for a given applied *force* to the structure. The relationship between the forces and corresponding displacements at the nodes of each element, *the element stiffness properties*, can be determined by virtue of its relatively simple shape. This is expressed as a set of simultaneous equations. The first step in the analysis is to represent these equations for each element, in matrix form, giving the elemental stiffness matrices. Using these individual stiffness matrices and associated transformation matrices, the overall (or structure) stiffness matrix for the entire structure is then assembled.

As there will be continuity of displacements and forces at all nodes in the structure this will give the following matrix equation:

$$\{F\} = [K] \{\delta\} \qquad\qquad (6.1)$$

where $[K]$ is the overall stiffness matrix of the structure. The overall force vector $\{F\}$ lists the externally applied forces at all the nodes, while $\{\delta\}$ is a list of nodal displacements. $\{F\}$ and $[K]$ are known and the objective is to find $\{\delta\}$.

Equation (6.1) shows that $[K]$ represents the force required to produce unit displacements of the structure. Therefore, if the model is

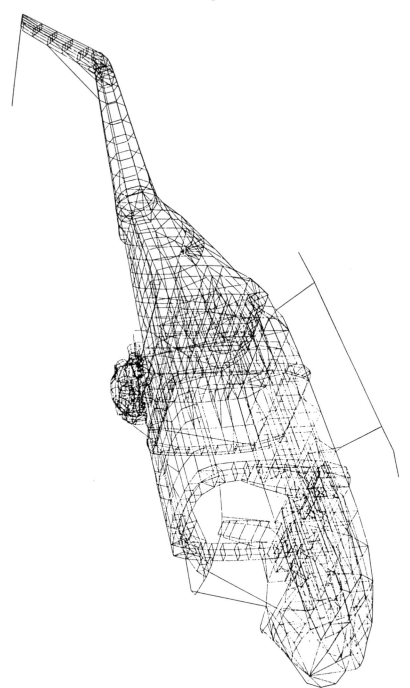

Fig. 6.3a. Finite element mesh of a Westland Lynx naval helicopter body, from NASTRAN finite element analysis (courtesy of Westland Helicopters Limited).

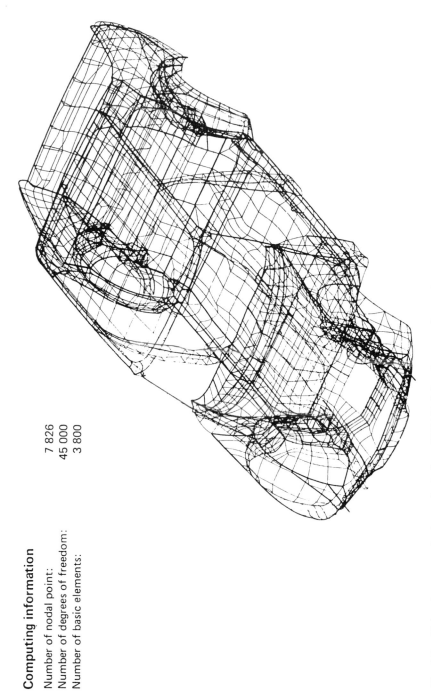

Computing information

Number of nodal point: 7 826
Number of degrees of freedom: 45 000
Number of basic elements: 3 800

Fig. 6.3b. Finite element mesh of a car body from SESAM finite element software (courtesy of Saab-Scania, Sweden).

considered as an equivalent spring, then $[K]$ will be a spring constant representing its stiffness.

Thus, the finite element method presented here is essentially one in which the analysis of a structure is carried out from the point of view of its stiffness. This concept is discussed in section 6.3.

With additional criteria which fully specify the problem to be analyzed (restraints, material properties etc.), and the known forces applied, the overall stiffness matrix equation can be solved for nodal displacements $\{\delta\}$, using numerical techniques usually based on *Gaussian elimination*. From these nodal displacement values, the strains and hence stresses are calculated for each node in the structure.

6.3   FINITE ELEMENT CALCULATION STAGES

To describe fully the mathematical principles of the finite element method would be beyond the scope of an introductory text of this nature. Instead, this section gives an overview of the calculation stages involved and an appreciation of the basic mathematical concepts in a finite element analysis. The user of a finite element package will be unaware of most of the stages. The loading and restraint conditions for a problem are usually known and the analysis is used to evaluate the strains.

Stages in the Analysis

1. The continuum is sub-divided (*discretized*) into an equivalent system of finite elements. The notation of a continuum or body is usually well defined. For example, an elasticity problem requires the sub-division of a deformable body. However, some continua are not so clear cut.

2. Selection of displacement functions: simple functions are chosen to approximate the distribution or variation of the displacements over each element in the structure. These functions are termed *displacement functions* or 'displacement models'. The unknown magnitudes of the displacement function are the displacements at the nodes (*or primary unknowns*).

   A displacement function is a linear approximation and can be expressed in various forms such as polynomials or trigonometric functions. The assumed displacement functions only represent an approximation of the exact displacement distribution.

3. Analysis of co-ordinate, property and topology data: the topology of

an element is simply a description of the arrangement of its associated nodes. The properties of a particular type of element depend upon the number and types of *degrees-of-freedom* and the basic assumptions made in deriving the stiffness properties. The degrees-of-freedom of an element refer to the nodal displacements, rotations and/or strains which are necessary to specify completely the deformation of the element. For example, the two-dimensional element shown in Fig. 6.4 would have 2 degrees-of-freedom per node.

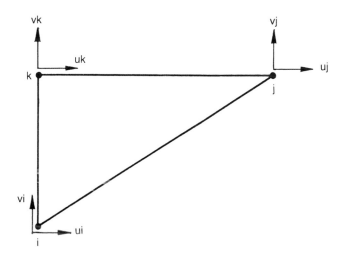

Fig. 6.4. Triangular two-dimensional in-plane element, showing the degrees-of-freedom associated with each node.

4. Derivation of individual element stiffness matrices: The stiffness matrix consists of the coefficients of the equilibrium equations derived from the geometric and material properties of an element and are obtained by a mechanical principle such as the minimization of potential energy.

   The potential energy of a loaded elastic body is represented by the sum of the internal energy stored as a result of the deformations and the potential energy of the external loads. If the body is in a state of equilibrium, this energy is a minimum.

   To illustrate this concept, consider the simple example of a rod in tension as shown in Fig. 6.5a. The rod has cross-sectional area A and Young's modulus E. The rod can be replaced by a linear spring as shown in Fig. 6.5b.

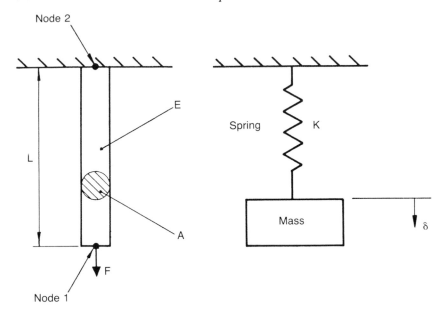

Fig. 6.5a. Simple rod in tension.   **Fig. 6.5b.** Spring analogy of rod in tension.

The spring having stiffness K is displaced by an amount δ due to a force, F.

The governing equation for a spring in tension is given by:

$$F = K\delta \tag{6.2}$$

The strain energy of the spring for displacements is:

$$E_\epsilon = \tfrac{1}{2}K\delta^2 \tag{6.3}$$

The potential energy E_p of the system, due to the displacement, has changed by $-F\delta$, so the system energy, E_s, is the addition of the strain energy and the potential energy, E_p, giving:

$$E_\mathrm{s} = E_\epsilon + E_\mathrm{p} \tag{6.4}$$

and to be in equilibrium this should be a minimum; hence:

$$\frac{\mathrm{d}(E_\mathrm{s})}{\mathrm{d}\delta} = 0$$

therefore $\dfrac{\mathrm{d}}{\mathrm{d}\delta}(\tfrac{1}{2}K\delta^2) + \dfrac{\mathrm{d}}{\mathrm{d}\delta}(-F\delta) = 0$ \hfill (6.5)

and $K\delta - F = 0$, then $\delta = F/K$ for equilibrium.

To make the calculation for this stage simpler, it is possible to select interpolation functions as the basis for the displacement model. The interpolation functions, or *shape functions* as they are better known, are functions with the property that their value is 1 at their own node and 0 at every other node.

The stiffness relates nodal displacements to nodal forces. The distributed forces applied to the structure are converted into equivalent concentrated forces at the nodes. The relationship between the stiffness matrix, nodal force vector and displacement vector can be expressed as a set of linear equations as in equation (6.1).

The elements of the stiffness matrix are regarded as influence coefficients. The stiffness matrix for the simple example shown in Fig. 6.5a can be evaluated by directly calculating influence coefficients. For example, if a downward force is applied to node 1, and node 2 is constrained, then a force, or influence coefficient, equal to AE/L is induced at node 1 and a force of $-AE/L$ is induced at node 2. The stiffness matrix for the element then becomes:

$$[K] = \begin{bmatrix} \dfrac{AE}{L} & -\dfrac{AE}{L} \\ -\dfrac{AE}{L} & \dfrac{AE}{L} \end{bmatrix} = \dfrac{AE}{L}\begin{bmatrix} 1 & -1 \\ -1 & 1 \end{bmatrix}$$

and equation (6.1) can be re-written as:

$$\begin{Bmatrix} F_1 \\ F_2 \end{Bmatrix} = \frac{AE}{L}\begin{bmatrix} 1 & -1 \\ -1 & 1 \end{bmatrix}\begin{Bmatrix} \delta_1 \\ \delta_2 \end{Bmatrix} \tag{6.6}$$

where δ_1 and δ_2 are the displacements at nodes 1 and 2, and F_1 and F_2 are the applied loads at nodes 1 and 2 (F_2 being zero).

5. Assembly of the overall system matrix equation: Individual element matrices for the whole continuum are merged to form a *global stiffness matrix;* this is followed by the formulation of geometric and boundary conditions. This merging process is known as *assembly* and is a common factor between all finite element analyses. The most widely used assembly technique is known as the direct stiffness method and requires the displacements at a node to be the same for all elements attached to the node.

6. Solution of primary unknowns: The global stiffness matrix equation for the system is solved. This is usually carried out by a reduction and

back substitution, which is a standard technique for solving simultaneous equations. In some finite element jobs the stiffness matrix is
very large (1000s×1000s). The result from this stage yields values for
primary unknowns (displacements) at each node in the structure.

7. Calculation of element strains and stresses, based on nodal displacements: Calculation of the element strains and stresses, the *secondary
unknowns*, can now be carried out (if desired) using simple matrix
algebra, based on the engineering relationship:

$$\epsilon = \delta/L$$

Elemental stresses can be found from the nodal strains using the
relationship:

$$\{\delta\} = [E]\,\{\epsilon\} \qquad\qquad (6.7)$$

where $[E]$ is the elasticity matrix.

6.4   FINITE ELEMENT TYPES

A finite element is in itself a one-, two- or three-dimensional body.
Hence, a structure can be easily subdivided as each element is essentially a part of the whole. In addition, continuous elements provide a
natural representation of the original continuum. Moreover, the concept of two- and three-dimensional elements has permitted the finite
element method to be generalized and applied to non-structural problems.

In practice, the elements chosen for a particular problem will be
selected from a library of elements as the most appropriate to represent
the more important characteristics of the continuum. Most commercial
finite element systems offer a wide range of elements, selected for their
accuracy and appropriateness to the problem. The most commonly used
finite elements may be categorized as follows:

(i) *Framework Elements — Rod and Beam*

This type of element, shown in Fig. 6.6, is used to model truss
and space frame structures. Usually it caters for bending in two
principal directions and twisting about its shear centre.

(ii) *Two-dimensional Plane Stress/Strain Elements*

These elements are normally used for finding stresses and displacements in thin structures when there are no appreciable
stresses normal to the surface and when the bending stresses are
minimal, i.e. buckling is assumed not to occur.

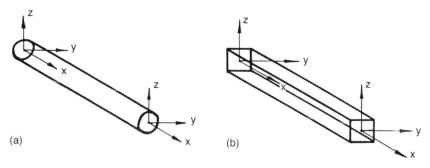

Fig. 6.6a. Simple rod framework element. **Fig. 6.6b.** Simple beam framework element.

The simplest of the plane stress/strain family of elements is the triangular element shown in Fig. 6.7a. Other types of two-dimensional elements in this group include rectangular and quadrilateral shapes (Figs 6.7b and 6.7c respectively). Many other geometric forms are possible in this class, but such other forms serve the more specialized purposes. The two-dimensional plane stress/strain element is probably the most widely used of all finite elements.

(iii) *Two-dimensional Plate Bending Elements*

This type of element is used for flat plates, shells and thin-walled members where the in-plane and out-of-plane effects are important (i.e. buckling). These elements tend to be of triangular or quadrilateral form (Figs 6.8a and 6.8b).

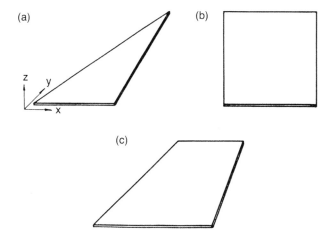

Fig. 6.7. Two-dimensional plane stress/strain elements.

(a)

(b)

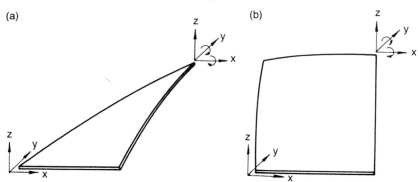

Fig. 6.8. Plate bending elements. These elements have rotational as well as translatory degrees-of-freedom.

(iv) *Axisymmetric Elements*

Many engineering components are *axisymmetric*, i.e. they are symmetrical about a centre line, and include: tanks, flywheels, rotors, shafts and pistons. The axisymmetric element is ideal for modelling such components. The elements, usually triangular or quadrilateral, are defined in a plane and then rotated about the plane, by an angle θ to form a solid. Figure 6.9 illustrates an axisymmetric triangular ring element.

(v) *Three-dimensional Elements*

Some engineering components are just too complicated for a two-dimensional analysis to solve with sufficient accuracy and therefore necessitate a three-dimensional solution. The most common three-dimensional elements are the *tetrahedron, rectangular prism* and the *hexahedron* (Figs 6.10a, 6.10b and 6.10c respectively).

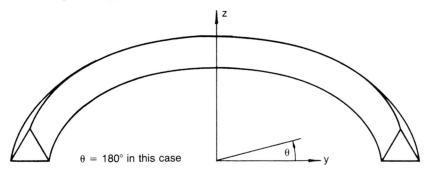

Fig. 6.9. Axisymmetric element.

(a)

(b)

(c)

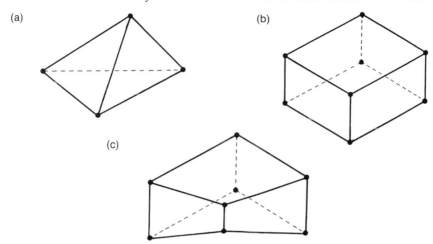

Fig. 6.10. Three-dimensional elements.
    a = Tetrahedral element.
    b = Rectangular prism element.
    c = Hexahedral element.

Three-dimensional finite element analysis is usually expensive due to the vastly increased number of calculations to be performed by the computer. Three-dimensional elements, therefore, tend to be used only when stresses vary in three dimensions or when a two-dimensional solution proves inadequate.

(vi) *Isoparametric Elements*

The final element to be considered here is the *isoparametric* type. Isoparametric elements are used when a curved boundary needs to be modelled. The elements achieve this curved boundary effect by having additional nodes positioned along the edges of the element and then constructing a line which joins the additional node with its two adjacent corner nodes. The new nodes are termed midside nodes and are usually equidistant from the corner nodes. Isoparametric elements can be used for two- and three-dimensional analysis. Figure 6.11 shows a selection of isoparametric elements.

6.5 MODELLING CONCEPTS

This section demonstrates some fundamental principles involved in the modelling of components by finite elements. Modelling principles are discussed by way of a simple example.

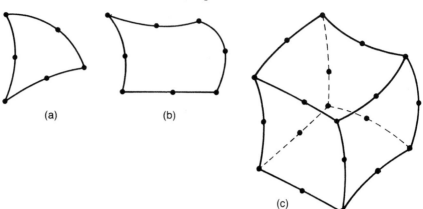

Fig. 6.11. Isoparametric elements.
        a = Triangular six-noded element.
        b = Eight-noded curvilinear, quadrilateral element.
        c = Twenty-noded three-dimensional brick element.

The Problem

Consider a flat steel plate subjected to direct tension, as shown in Fig. 6.12. The plate has a circular hole in its centre and the stresses in the vicinity of the hole are to be determined. The plate is symmetrical about the centre lines of the hole; therefore it will only be necessary to model one quarter section of the plate (see Fig. 6.13). This will dramatically reduce the number of elements required to model the component.

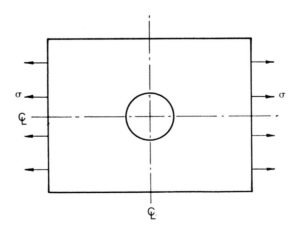

Fig. 6.12. Flat plate in tension.

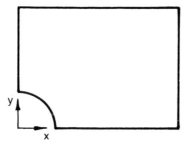

Fig. 6.13. Quarter section of original plate for modelling.

Choice of Element

In order to obtain reliable results from a finite element analysis, it is important to select elements which are most appropriate for the job. As this component is a flat plate, a two-dimensional approach will be adequate; therefore 2-D plane stress/strain elements will be used. All quadrilateral or all triangular elements could be employed but to demonstrate that mixed assemblages are possible, a combination of the two types of element will be used.

Isoparametric elements will be used to model satisfactorily the curved boundary at the section of the hole.

It is important when drawing up the mesh to take account of the shape of the elements, as distorted quadrilaterals and high aspect ratio triangles give unreliable results. Ideally the aspect ratio of a triangular element should be 1:1 and the ratio of the longest to shortest side of a quadrilateral element should not exceed 3:1 (Figs 6.14a and 6.14b respectively). Highly unequal-sided elements will lead to unreliable results.

Producing the Mesh

Having decided upon the types of element to use and the type of analysis to be carried out, the mesh can be produced. The section of the plate has to be sub-divided into an equivalent system of finite elements. This process is essentially an exercise of engineering skill and judgement. The elements have to be arranged in such a way that the original plate section is simulated as closely as possible. The general objective of this subdivision (or discretization) process is to divide the component into elements sufficiently small so that simple displacement models can adequately approximate the true solution.

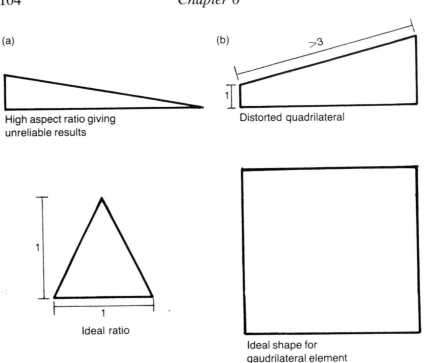

Fig. 6.14. Ideal element proportions for triangular and quadrilateral elements.

Figure 6.15 shows a first attempt at discretization of the plate section.

Refining the Mesh

Most engineering structures and components have zones in which pronounced variations in stress and strain occur, e.g. high stress gradients around keyways, sharp angles, corners and radii. The component under consideration here will have a high stress concentration in the area of the hole, and, as it is this region which is of interest, a higher order of accuracy of results is required here than for the remainder of the domain. It is possible to increase the accuracy of the solution by having a concentration of relatively small elements in selected regions. A refined mesh of the plate is shown in Fig. 6.16, illustrating the smaller elements in the high stress area. Isoparametric elements are used in the near vicinity of the hole to ensure the curved boundary is approximated as closely as possible. The application of this type of element makes it

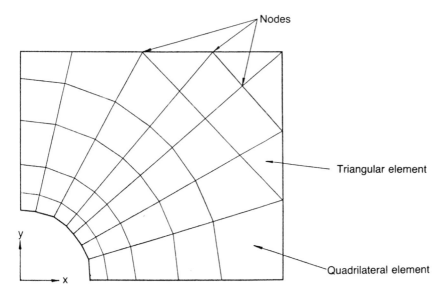

Fig. 6.15. First attempt at discretization of component. The figure shows the elements and nodal points which make up the mesh.

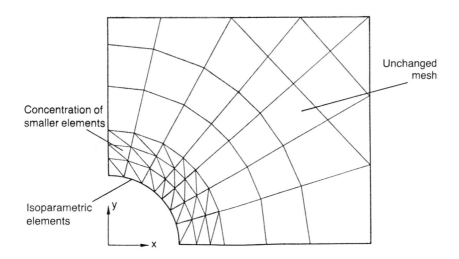

Fig. 6.16. Refined mesh showing the high concentration of small elements in the region of the hole.

possible to represent the hole more exactly than could be achieved using conventional two-dimensional elements.

It should be noted at this point that a finite element analysis will converge to an accurate solution of the problem if a larger number of relatively small elements are used. Each element in the mesh is assigned a number and similarly the nodes which are attached to the elements are numbered in a given sequence for subsequent data entry.

Boundary Conditions

Having finalized the mesh, it is now possible to apply *boundary conditions* to the model. The model will not be completely specified unless boundary conditions are prescribed. If these conditions are absent or insufficient, the element stiffness matrix and the overall stiffness matrix will be singular and the problem becomes mathematically insoluble. The boundary conditions represent the way in which the model is constrained in free space and the system of loads which are externally applied. The loading and restraint conditions for the plate model are given in Fig. 6.17. As the mesh is a representation of one quarter of the component, restraints are applied to satisfy the condition of a body which is symmetrical about two axes. Edge A in Fig. 6.17 is restrained from moving in the Y direction, thus simulating the full width of the plate. However, translation is allowed in the X direction, as this is the direction of loading. Similarly, edge B in Fig. 6.17 is prevented from

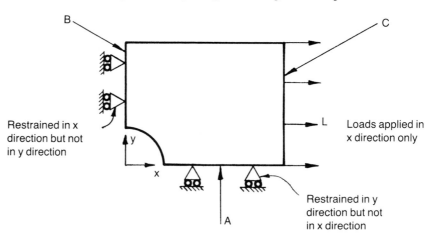

Fig. 6.17. The boundary conditions (the restraint parameters and loading conditions can be seen).

moving in the X direction, again to satisfy symmetry, but is allowed movement in the Y direction to cater for the Poisson effect of the material. Pre-determined external loads are applied to edge C, subjecting the plate to direct tension. In general, boundary conditions can be specified intuitively from a knowledge of the expected deformation of the structure under load.

The way in which the restraints and loads are expressed for use by an FE model partly depends upon the software package used and the element type. In most cases, restraints and loads are applied to individually specified nodes although many FE systems enable these boundary conditions to be applied to planes or lines of nodes.

From the point of view of model preparation, the enginering skill and judgement is now complete. All that remains is to compile a data file describing the model in terms of nodal co-ordinates, element topologies, boundary conditions and so on. Then allow suitable FE software to read the data, compute the solutions and print the results of analysis. The engineer can then interpret the results and make whatever decisions are required.

6.6   FINITE ELEMENT SOFTWARE

By the early 1970s, several general-purpose finite element analysis computer programs had been developed. These programs concentrated on providing the latest FE theory and the latest matrix solution techniques in general analysis formats. This pre-occupation with technique rather than user-friendliness was well founded, since during the early life of the programs, substantial advances were still being made in the analytical content of the method, as well as the facilities available on the computer hardware on which the programs were intended to run.

Presentation of the code was also less important as the application of the code was still confined mainly to research and development environments. However, today the finite element method has reached the majority of engineering disciplines, partly due to commercial exploitation of large software packages. Whilst progress in introducing the finite element method has been rapid, attempts at improving the usability of programs have been piecemeal.

The engineer wishing to carry out an analysis quickly found that decisions on idealization, the manual work involved in representing the component under analysis, preparing the data for program code, checking the data and interpreting the results could be a formidable task.

To alleviate these problems, pre- and post-processor programs were

developed and built around the analysis systems to reduce the tasks of
data generation and checking by providing automatic and semi-automatic generation of regular parts of a mesh and graphically displaying
results of the analysis.

Pre-Processor

A pre-processor program is used to define, verify and format data from
a finite element mesh. This data will usually consist of nodal points,
elements, material properties, restraints and additional miscellaneous
data. The pre-processor arranges the data into a specified format for
acceptance by the analysis program, as shown diagrammatically in Fig.
6.18.

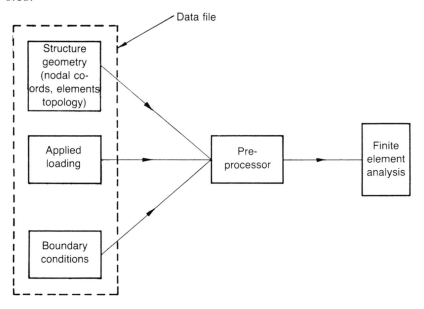

Fig. 6.18. Diagrammatic representation of how the pre-processor interfaces with the
analysis software.

Post-Processors

A post-processor program is used to determine the significance of the
finite element analysis by producing graphic displays and documentation of the results to supplement or replace traditional computer print-
outs. Post-processor programs interface with the finite element analysis,

as shown in Fig. 6.19, and are particularly useful for organizing, manipulating and outputting results for those areas in which the engineer is currently interested. Post-processors can, for example, carry out the following:

(i) Search for values outside a given range, i.e. displacements above a prescribed value.

(ii) Highlight stresses of a particular nature (tensile or compressive) in selected regions.

(iii) Sort results into a specified order, e.g. according to a thermally induced stress value.

The majority of pre- and post-processors are designed to interface to a variety of finite element analysis software.

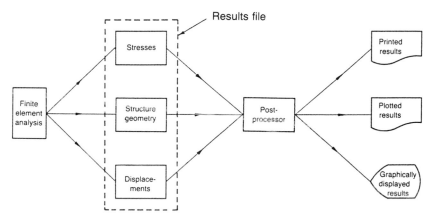

Fig. 6.19. Diagrammatic representation of how the analysis software interfaces with the post-processor.

Output of Results

The output of results obtained at the end of an analysis varies from program to program, but, in general, most programs will yield the following basic output:

(i) A list of displacements at each node in the structure.

(ii) A list of the forces at each node, broken down into the contributions from each element.

(iii) Stresses in global co-ordinates for each element.

(iv) Principal stresses and maximum and minimum shear stresses and the directions in which they act.

Figure 6.20 shows, diagrammatically, the stress and displacement output for a single node in a two-dimensional analysis. A visual representation of the results is provided by most software packages and includes: displaced shapes (superimposed on original mesh if required), stress contour and stress vector plots.

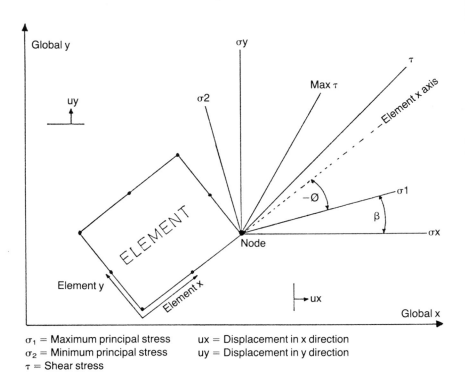

σ_1 = Maximum principal stress   ux = Displacement in x direction
σ_2 = Minimum principal stress   uy = Displacement in y direction
τ = Shear stress

Fig. 6.20. Nodal stress and displacement system. An element may have its own axis set independent of the global axis.

Interpretation of the results is sometimes required in view of the limitations of the model and/or elements used. This usually amounts to simple *interpolation* or *extrapolation* of results to overcome the shortcomings of a finite mesh size, particularly with elements such as the plane stress triangle. As this element is widely used, an indication of the method of interpreting results obtained with this element will be useful.

The plane stress/strain triangular element is based on the assumption of linear displacements between nodes for the derivation of its stiffness properties. Consequently, the element can only represent a constant stress or strain in each of its co-ordinate directions. It is usual, therefore, to allocate this stress or strain to the centroid of the element. A plot along a given line in the structure can be made by joining the centroid values as in Fig. 6.21. Values between the centroids of the elements are then interpolated as required. The stresses at the nodes can be estimated by averaging the values obtained for the elements surrounding the node. The use of both centroidal and nodal stresses together is usually sufficient to give an adequate picture of the stress field in a structure, providing that a sufficiently fine mesh has been used. This averaging and interpolating process is applicable to many of the simpler elements and, in general, an intuitive approach to the interpretation of finite element results, for the simpler elements is often the correct one and reduces the need for a finer mesh.

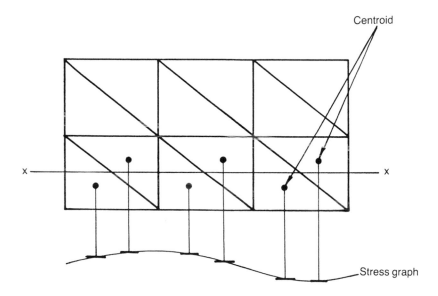

Fig. 6.21. Representation of centroidal stress, giving stress averaging along a line.

State of the Art

Advances in computer hardware and software have increased very rapidly during the last few years and of particular importance to FE is the widespread availability of: (i) low cost computer terminals, providing local computing as well as access to mainframe; (ii) remote computer networks offering a range of powerful analysis facilities; (iii) less expensive graphics hardware and software enabling a wider adoption of interactive graphics for engineering design; and (iv) database management techniques for easy updating and retrieval of data.

Finite element software, in particular, is at such an advanced state of development that complexity of component shape is no longer a problem, as libraries of elements from simple beams to solid bricks are available, allowing mathematical modelling of the most complex geometry.

The most complicated loading patterns, inexpedient pressure distribution, and thermal and dynamic loading can be readily analysed. Features for *anisotropic, orthotropic* and special materials, such as glass reinforced plastics, are commonplace.

No longer are engineers confined to static analysis. They can now calculate heat flow through a component, its natural frequencies, and how and where it will deform; they may even subject it to shear loads, physical or thermal, and examine its response. It is also possible to calculate *creep* and examine *non-linear* behaviour.

There are, however, problems in practice, not least of all cost. The more complicated the model, the more elements are needed to model it, hence more computer time is required. Complicated models place heavy demands on the engineer's time and are prone to data errors. Cost can also be a limiting factor in more advanced analyses, where the model is fairly simple but has to be run many times to simulate time histories.

Another drawback is the need to specify the model numerically, e.g. all material properties and loading must be definable in pattern and magnitude.

Types of Software

One problem, which anyone investigating finite elements will find, is the vast array of FE software available on the market. All of the major packages work on a similar basis, and in many cases use common element types. Some packages reflect their origins; for example, programs developed by power industries have specialized axisymmetric elements, while aerospace orientated programs have good shell elements.

One difference between finite element programs is the amount of pre- and post-processing they provide. The data input to an FE program can be considerable, the output also large. Programs are available as part of a standard package, or independent, to make the creation and interpretation of models both faster and more reliable. The independent, or stand-alone programs interface with a variety of different packages for different purposes which may be very useful.

Essentially there are *three* ways of using finite element software. One is to buy or lease the software and mount it in-house on one's own computer. This may require someone employed on a full-time basis to support the user software and the extra demands on the computer may mean an upgrade or a new machine. The second option would be to use a bureau. If a bureau is chosen the user may have the whole job carried out or may prefer to create his/her own data file and submit this for analysis. There are some distinct advantages in using a bureau — namely a wide choice of analysis package is available, the need to procure extra hardware is eliminated, the demands on in-house software support are lessened and the use of a large machine often produces rapid turnround. Bureau staff will be more familiar with packages than will individual company staff.

The third method is to sub-contract the work, i.e. paying specialists to carry out all or part of the analysis. Many bureaux and software companies now offer this service, and for the occasional user of finite elements this method proves most cost effective.

Automatic Meshing and Graphic Analysis

Data preparation and results interpretation are the two most time-consuming aspects of finite element analysis, thus any attempts to reduce this time is an important factor in any FE job. Most of the large finite element suites now incorporate interactive graphics systems to provide for automatic mesh generation and graphical results analysis.

Systems are available which enable the user to develop a mesh interactively. Part of the structure is drawn on the screen then a suitable mesh is produced, automatically, throughout that area. The mesh may be modified or duplicated in various ways to represent accurately the structure to be analyzed. To view post analysis results, systems may be used to provide colour contours of stress or temperature or to show dynamically the displaced shape of the model. Graphic output is an easier way of assessing a large amount of computer output.

Figure 6.22 shows colour temperature contours drawn by an interactive graphics FE program.

Another example of a graphical aid to analysis is the type of program capable of hidden line removal. This feature gives uninhibited representation of complex structures, as shown in Fig. 6.23.

Chapter 6

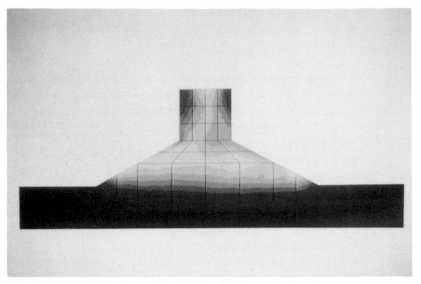

Fig. 6.22. PIGS (Pafec Interactive Graphics Suite) output, showing temperature contour shading on a finite element mesh (courtesy of PAFEC Limited).

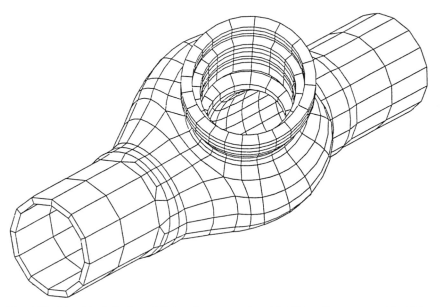

Fig. 6.23. Graphical display of finite element mesh, with hidden lines removed by IMAGE processor (courtesy of Sulzer Bros Switzerland).

Standardizing the Software

There now exists a large number of different finite element packages on the market, the majority of which purport to offer solutions to nearly every conceivable structural problem. The packages contain vast amounts of computer code and the type of code differs from one package to the next. There has been a certain amount of unease recently, particularly within the nuclear and aerospace industries, with regard to the accuracy of results of some FE analyses and the implications of these results.

The general feeling in some quarters was that a tightening-up of procedures and standardization of code was required. To that end, a body has been set up, NAFEMS, in an attempt to improve the standard of finite element programs. NAFEMS (National Agency for Finite Element Methods and Standards) was set up by the National Engineering Laboratory with funding from the Department of Trade and Industry's Mechanical and Electrical Engineering Research Requirements Board.

6.7   APPLICATION OF THE FINITE ELEMENT METHOD

As stated earlier, the finite element method can be used for a plethora of engineering applications which would be impossible to cover in this text. However, as an introduction to the application of the method, this section describes how one particular software package operates, and the two sections which follow demonstrate the application of the software to the solution of two very different engineering problems.

The PAFEC System

PAFEC stands for Program for Automatic Finite Element Calculation and is the name of both the company and its finite element system. Work on the program began in 1964, when research staff at Nottingham University started to investigate the method. The success of the program resulted in seven of the staff setting up the company in 1976 to market it.

The following is a brief introduction to the workings of the Pafec system. The software is configured to run in ten sections, called *phases* within the package. A given job will run some, but not necessarily all, of the phases, although phases 1, 4, 6 and 7 are the minimum requirements for the calculation of primary unknowns. The main function of each phase is given in Table 6.1.

Table 6.1. Synopsis of operations for each of the PAFEC phases.

PHASE	Short description	Detailed description
1	Read	Data modules are read in, default values are inserted and the modules are placed onto backing store. The NODES module is expanded so that all mid-side nodes are included.
2	PAFBLOCKS	Any PAFBLOCK data is replaced by the full nodal coordinate and topological description of the complete mesh elements.
3	IN.DRAW structure	The structure itself is drawn. At this stage it is not possible to show any results such as displacements, stresses or temperatures, since these have not yet been evaluated.
4	Pre-solution housekeeping	In this PHASE the constraints on the problem are considered and a numbering system for the degrees of freedom is derived.
5	IN.DRAW constraints	This PHASE is very similar to the PHASE 3 except the constraints which have been applied are shown. Conversely the degrees of freedom can be indicated on a drawing.
6	Elements	The stiffness (or other such as conductivity, mass etc) matrices of all the elements are found and put onto backing store.
7	Solution	The system equations are solved for displacements, temperatures, or whatever happens to be the primary unknowns in the problem being tackled.
8	OUT.DRAW displacements	The primary unknowns in the problem (i.e. displacements or temperatures) are drawn.
9	STRESS	The stresses are found.
10	OUT.DRAW	Stress contour, stress vector plots etc. are produced.

The need to execute a particular phase and the type of calculation to be carried out is determined by PAFEC from the data supplied by the user. The path taken by the program to form the solution to the given problem can be controlled by the user by means of a *CONTROL* module. This module is used to specify three categories of information.

(i) Statement of which phases are to be executed in a run.

(ii) Phase-dependent specifications, defining values or actions which relate to a particular phase.

(iii) Information defining the type of problem to be solved.

Pafec Data

A Pafec data file contains a series of data modules describing the specific details of the problem, such as Nodes, Elements, Loads, Material, and ending with an END. OF. DATA instruction.

An example module is shown below:

NODES ←——— module header

$Z = 4.7$ ←——— constant property card (i.e. all these nodes on a constant plane)

NODE. NUMBER X Y ←——— contents card

17, 14.63, 3.85
18, 15.66, 4.975   Table of data (i.e. node No. x co-ordinate y,
19, 15.66, 12.25   co-ordinate)
213, 25.45, 19.7

The data file, containing control module and data modules, is then used as input to the Pafec suite.

6.8   CASE STUDY — CURVED CANTILEVER

The Problem

It is required to find the stress distribution in a curved cantilever which has two point loads acting at its free end, as shown in Fig. 6.24.

Input

To model this structure adequately about 25 elements would be required, which would necessitate the manual description of each element and a minimum of 96 nodes. However, the entire mesh of 25 elements can be generated using only 6 manually defined nodes describing one block of elements. This block is termed a PAFBLOCK and can be regarded as a semi-automatic mesh generation tool.

Overleaf is the complete data file for this job. A 'C' in the first column indicates a comment card and is used here to make the file self-explanatory.

The x and y co-ordinate terminology in this example relate to r,θ polar co-ordinates respectively.

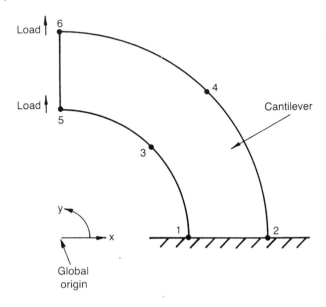

Fig. 6.24. Curved cantilever with point loads, restrained at its lower edge.

DATA FILE

```
CONTROL
CONTROL.END
C  A blank control module means use default options
TITLE     CURVED CANTILEVER UNDER POINT LOAD

NODES
AXIS=3
NODE X Y
C  Although this module has x,y co-ordinate headings, the values are
C  actually r,θ polar co-ordinates
1     4 0
2     6 0
3     4 45
4     6 45
5     4 90
6     6 90
C  This module gives the coordinates of the nodes.

MATERIAL
MATERIAL.NUMBER        E        NU
1                     3E06      0.3
```

C There are ten standard materials programmed into Pafec which may
C be overwritten or new ones may be used.
C The entries give respectively; the material number, Youngs modulus
C (in appropriate units) and Poissons ratio.
C Additional material properties may be included to fully describe
C the material(s) to be used.

PAFBLOCKS

N1	N2	ELEMENT.TYPE	TOPOLOGY
1	1	36210	1 2 5 6 0 3 4

MESH

REFERENCE	SPACING.LIST
1	5

C N1 and N2 are the direction of the Pafblock module to the
C appropriate spacing list (ie 5 by 5 elements) in the mesh module.
C 36210 indicates the element type and in this case refers to an
C eight noded isoparametric curvilinear quadrilateral.
C Topology defines the nodes on the Pafblock

LOADS

NODE.NUMBER	DIRECTION.OF.LOAD	VALUE
5	2	1500
6	2	1500

C Apply 1500 force units (default Newtons) each to nodes 5 and 6
C in the global y direction

RESTRAINTS

NODE.NUMBER	PLANE	DIRECTION
1	2	0

C Constrain node 1 and all other nodes which lie on a constant y
C plane. Zero direction means all degrees of freedom at these nodes
C to be restrained.

STRESS.ELEMENT

START	FINISH	STEP
1	25	1

C The stress.element nodule is used to define which elements in the
C mesh require stressing (in this case all 25). This module is
C useful if only selected elements in a model are to be stressed.

IN.DRAW

TYPE

2

C The in.draw module is used to create a drawing of the completed mesh
C and is useful for error checking. Many options are available within
C this module. Type 2 gives the minimum information required.

OUT.DRAW

PLOT.TYPE

1

30

C This module describes drawings of the results of calculations.
C Plot.type takes the following values: 1=displaced shape, 30=contours
C of maximum in-plane principal stress.

END.OF.DATA

The completed data file is given a name and submitted to the Pafec program.

Output of Results

The results from a Pafec run can amount to a considerable quantity of information, especially if most of the phases have been employed or if the model comprises a large number of elements. Therefore, whilst the output from the main phases will be described, only the graphical results will be given.

Phase 2. Pafblock Generation Phase

Output from this phase gives the global Cartesian co-ordinates of pre-defined and self-generated nodes and the topology data to show the relationship between the elements and their respective nodes for the whole structure.

Phase 3. Mesh Drawing

Figure 6.25 gives the complete IN.DRAW plot, clearly showing the organization of the elements.

Phase 4. Constraints and Degrees of Freedom Data

This output gives constraint and loading data. The details in this phase may be used to check that constraint data is correct.

Phase 7. Displacements

This output gives the displacements in appropriate units for a selection

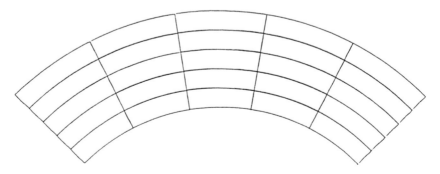

Fig. 6.25. Finite element mesh of cantilever produced with the aid of the PAFBLOCKS module.

or all of the nodes in the structure. Constrained nodes can also be highlighted. Again, these values can be printed if desired.

Phase 8. Displaced Shape Plot

The displaced shape, superimposed on the original mesh, is shown in Fig. 6.26.

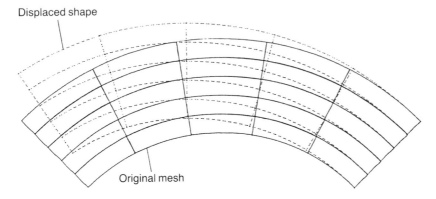

Fig. 6.26. Displaced shape plot. The loaded cantilever in its displaced position is shown dotted and is superimposed on the original mesh.

Phase 9. Stresses

A comprehensive stress analysis can be printed out in tabular form.

Phase 10. Stress Contours

The Phase 10 stress contour plot is given in Fig. 6.27.

6.9   CASE STUDY — STRUCTURE NATURAL FREQUENCIES

The Problem

To determine the natural frequencies in a restrained three-dimensional structure made from two different materials (Fig. 6.28).

Input

This structure is to be modelled using four elements, as shown in Fig. 6.29. The pafblocks facility could be used here, but to illustrate use of the *elements* module the elements are to be defined individually. The fully commented data file is given on pages 123–124.

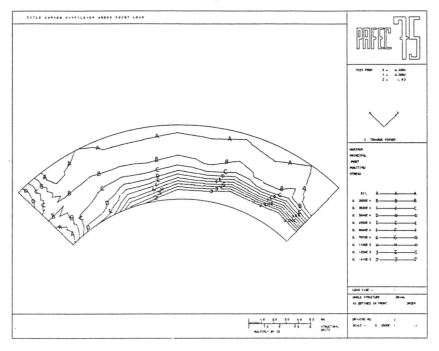

Fig. 6.27. Stress contour plot of cantilever. The figures on the right indicate the maximum principal stress.

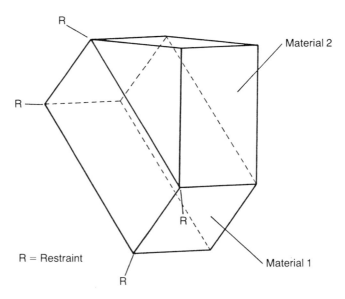

Fig. 6.28. Three-dimensional body, comprising two materials, for vibration analysis.

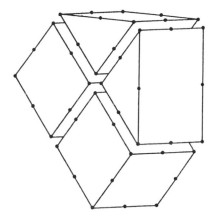

Fig. 6.29. View of four three-dimensional elements chosen for analysis.

DATA FILE

CONTROL
CONTROL.END
TITLE STRUCTURE NATURAL FREQUENCIES
C An abbreviated form of data input will be given for the nodes
NODES
NODE.NUMBER X Y Z
1 0 1 0//2 1 1 0//3 2 1 0//4 3 1 0//5 4 1 0//6 0 2 0//7 2 2 0//8 4 2 0
9 0 3 0//10 1 3 0//11 2 3 0//12 3 3 0//13 4 3 0//14 0 1 1//15 2 1 1
16 4 1 1//17 0 3 1//18 2 3 1//19 4 3 1//20 0 1 2//21 1 1 2//22 2 1 2
23 3 1 2//24 4 1 2//25 0 2 2//26 2 2 2//27 4 2 2//28 0 3 2//29 1 3 2
30 2 3 2//31 3 3 2//32 4 3 2//33 1 0 0//34 2 0 0//35 3 0 0//36 2 −1 0
37 2 −1 1//38 1 0 2//39 2 0 2//40 3 0 2//41 2 −1 2
C Multi records may be given on a single line
ELEMENTS
NUMBER GROUP ELEMENT.TYPE PROPERTIES TOPOLOGY
1 1 37110 1 1 3 9 11 20 22 28 30 2 6 7 10 14 15 17 18 21 25 26 29
2 1 37110 1 3 5 11 13 22 24 30 32 4 7 8 12 15 16 18 19 23 26 27 31
3 1 37210 2 1 36 3 20 41 22 33 34 2 14 37 15 38 39 21
4 1 37210 2 36 5 3 41 24 22 35 4 34 37 16 15 40 23 39
C Element type 37110 is a twenty noded isoparametric brick with six
C curvilinear faces and twelve edges.
C There are eight corner nodes and 1 mid-side node on each of the edges.
C Element type 37210 is a triangular prism element having fifteen
C nodes. The properties entry relates to the material properties in
C materials module.
MATERIAL

MATERIAL	E	NU	RO	ALPHA
1	30.0E+06	0.30	0.00065	11.0E−06
2	30.0E+03	0.30	0.00065	11.0E−06

C Ro and Alpha give respectively the mass density and coefficient of
C linear thermal expansion of the materials.
RESTRAINTS

NODE.NUMBER	PLANE	AXIS.NUMBER	DIRECTION
1	0	0	123
5	0	0	123
9	0	0	123
13	0	0	123

C Nodes 1 5 9 and 13 are restrained in the x,y and z translatory
C directions.
LOCAL.DIRECTIONS

NODE.NUMBER	LOCAL.AXIS
3	4
7	4
11	4

C Local.directions are used here in conjunction with the axis module
C to define a plane which does not lie on the global x y or z plane.
C It is used to ensure that displacements on sloping planes are
C in the correct direction.
AXES

AXISNO	RELAXISNO	TYPE	NODENO	ANG1	ANG2	ANG3
4	1	1	1	0	−30	0

MODES.AND.FREQUENCIES

AUTOMATIC.MASTERS	MODES
30	3

C The modes.and.frequencies module is used to define the type of
C dynamic analysis to be carried out. Automatic masters is the number
C of automatic master degrees of freedom that the system is to choose
C in addition to the users masters. The modes entry indicates the
C number of modes of vibration to consider (in this case 3).
IN.DRAW

TYPE.NUMBER	INFORMATION.NUMBER
3	23

C Information number 23 means display node and element numbers on
C input (phase3) drawing.
OUT.DRAW

PLOT.TYPE	CASE.NUMBER
1	1
1	2
1	3

C The case entries here describe the modes of vibration to be
C represented in the output drawings.
END.OF.DATA

Output of Results

Here again, the following information is an edited reproduction of the
output for selected phases of the problem.

Phase 3. Mesh Drawing

Figure 6.30 gives the IN.DRAW plot. The outline of the structure is shown as a continuous line with the element division shown dotted. (Node numbers are encircled.)

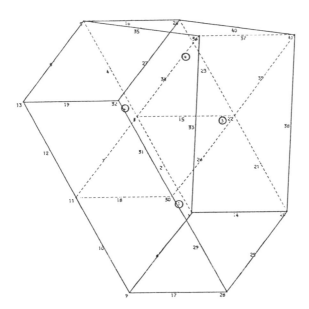

Fig. 6.30. Drawing of mesh showing element and nodal numbering.

Phase 7. Natural Frequency Displacements

This output gives the natural frequencies at each mode shape together with the nodal displacements.

Figures 6.31a, 6.31b and 6.31c show the mode shape for three modes of vibration at the frequency given in the respective diagram.

6.10   BENEFITS OF THE FINITE ELEMENT METHOD

The benefits of the finite element method and the contribution it makes to Computer-Aided Design may already have become apparent. However, a summary of the major advantages and limitations of finite elements is given overleaf:

(a)

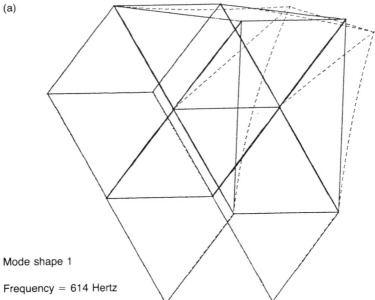

Mode shape 1

Frequency = 614 Hertz

(b)

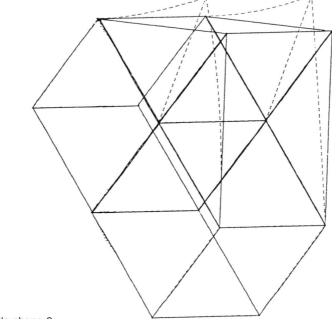

Mode shape 2

Frequency = 738.4 Hertz

(c)

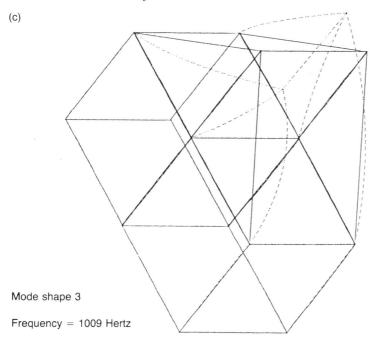

Mode shape 3

Frequency = 1009 Hertz

Fig. 6.31. Mode shapes shown dotted for three modes of vibration at the frequencies stated on each diagram.

Advantages

Engineering components possessing complex shapes can be modelled easily for analysis by dividing the continuum into small bodies. The fact that the boundary conditions are easily alterable means that a great variety of operating regimes and conditions may be applied to the model.

Not only can the FE method cater for complex assemblies and boundary conditions, but it is also capable of representing nonlinear, non-homogeneous and composite matrials, and performing types of analysis which would otherwise prove difficult in practice using a conventional analytical approach.

The use of finite elements coupled with the power of modern-day computers enables design solutions to be reached with speed and accuracy, making it possible to reduce design lead-times and to predict likely operational problems. The accuracy of finite element results compares well with, and is often better than, results from other analytical methods or experiments on protypes or models.

If an automatic mesh generation system is employed, the user will be able to see the element configuration as the mesh generation progresses. Moreover, the mesh can be changed, interactively, to arrive at an *optimum* idealization of the structure.

As the finite element method can be used for simulation and prediction of problems, it *reduces* the need for exhaustive field testing resulting in an overall reduction in costs. It also eliminates the propensity to over-design to compensate for ignorance of service stress thresholds.

More recently it has been possible to integrate the finite element method into a total CAD/CAM system, including design, draughting and manufacture of a product. In this way, one common database is used for all aspects of a given component, eliminating duplication of FE and other design/manufacture data and thus realizing enormous savings in time and cost. This also allows easy re-design if the results from the FE analysis are unfavourable.

Limitations

It would be naive to suggest that the application of the finite method did not suffer from drawbacks. There are, of course, limitations with the method but these tend to be overshadowed by the obvious benefits incurred.

The limitations may be summarized as follows:

Even the most efficient finite element software requires a large amount of computer space and time to run. Hence, the method is, to a large extent, limited to users with access to powerful 16- or 32-bit computers with their attendant time sharing, remote batch and virtual memory capabilities. However, some FE packages are specially designed to run on desk-top computers, in which case the computer tends to operate in a dedicated manner.

In view of the size of finite element packages (which often exceeds 100,000 lines of source program), and in spite of rigorous testing methods, there is always a possibility of program error.

Data preparation is one of the most tedious aspects of a finite element analysis and can be a drain on precious man-hours if mesh generation aids are unavailable.

There are some tasks for which the accuracy of solutions obtained are difficult to predict and for which finite element modelling proves difficult — for example, the representation of contact behaviour between two independent bodies.

Engineering judgement must always prevail in the initiation of a finite element model and certain critical parameters cannot yet be left to the judgement of a mesh generator, as only a person with sound engineering knowledge of the behaviour of a given range of components would be able to satisfactorily specify the mesh for analysis. Similarly the interpretation of results has to be carried out with care and any assumptions used in the formulation must be borne in mind.

Chapter 7

Computer-Aided Manufacture

7.1 INTRODUCTION

Whilst Computer-Aided Design (including draughting, modelling and analysis), is an important and integral part of Computer-Aided Engineering, the ultimate aim for the majority of engineering companies is to manufacture and consequently market their product.

This chapter gives the background of Computer-Aided Manufacture and explains its role in the CAE process by describing generally the important activities within computer-aided manufacturing. The following two chapters give more specific details on two aspects of manufacture, i.e. machine tool control and engineering robotics.

For many years manufacturing industries followed relatively unchanged procedures and practices in order to produce their engineering products. The basic elements were the drawing office, producing detailed drawings, parts lists and bills of material; and the machine, assembly, fabrication and fitting shops, producing components from paper specifications and instructions. However, during the 1960s it became evident that advances in computer technology could be effectively applied to engineering manufacture. One factor which had a great bearing on the introduction of computers to manufacturing was the rapidly developing techniques used in the numerical control of machine tools.

It soon became clear that the potential advantages were not restricted to machine tools and with the partial integration of Numerical Control with Computer-Aided Design, companies were able to increase competitiveness; by raising productivity, maximizing production capacity; minimizing lead-time, and enhancing product quality and consistency. Other important commercial benefits included: meeting competitive delivery dates; eliminating delay on repeat orders; improving cash flow; reducing overheads; providing management information to assist corporate planning and to reduce work in progress. Manufacturers are increasingly regarding computers as representing significant potential for increasing the productivity of the company, and computer system vendors are seeing the manufacturing market as representing more potential business than any other market. It was evident during a recent visit to a UK machine tool exhibition that over 90% of all the exhibits were, to some extent, computer controlled.

Computers are steadily taking over total control in manufacturing environments and it may not be too long before the totally unmanned factory becomes common place.

The development of computer-aided manufacturing systems has, to some extent, been fragmented, with numerical control and computer numerical control developing independently of industrial robot research, although now there are increasing attempts to combine these two key areas to provide a complete manufacturing system.

The shortage of skilled labour in certain areas of production has encouraged companies to investigate computer methods as a means of overcoming production problems. This has resulted in the merging of CAD and CAM to provide a truly integrated manufacturing system.

7.2 LET THE COMPUTER DO THE WORK

Although some traditional manufacturing tasks require the operator to have a high degree of skill and knowledge, there are many jobs which can be tedious, time consuming or environmentally unpleasant.

The computer can be used in manufacture to free the operator from the machine, enabling him/her to concentrate on more challenging and creative work. There are many aspects of manufacture where the computer can play a key role, including machine tools, robotics, quality control, inspection and testing, factory automation and manufacturing management.

Figure 7.1 shows how the various elements of computer-aided manufacture interrelate.

7.3 COMPUTER NUMERICAL CONTROL

Numerical control is a method of controlling a machine by providing it with numerical information. It has been possible to control machines in this way for over 100 years by having the program of instructions permanently fixed on a drum or similar rotating device. However, the early control systems, by virtue of their cyclic operation, were restricted to the manufacture of components made up of straight lines and circles. Then, in the early 1950s, machines were fed with information by means of a punched paper tape, which meant that the program no longer had to be cyclic and much more flexibility could be built in to the machines'

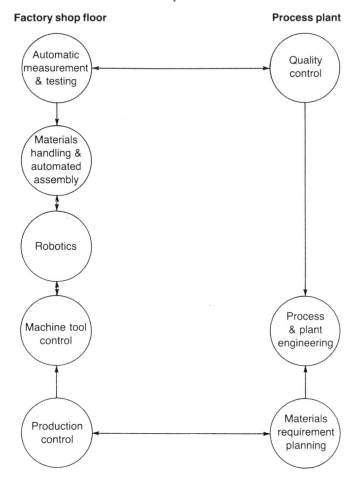

Fig. 7.1. Interrelation of CAM activities.

capability.

Special programming languages were developed to control the machine tools, one of which was the *Automatically Programmed Tools* (APT) system. APT was developed at the Massachusetts Institute of Technology and is a program language allowing geometrical co-ordinate data and tool motion statements to be specified for any numerical control machine. APT allows the user to produce three-dimensional geometrical tool contours of tool motion using simple English-Orientated statements. The objective of the APT system is to reduce the task of calculating relative motions between tool and workpiece to produce a given shape. It can be used to produce control tapes for simple and complex parts.

The implementation of NC machines in the machine tool industries grew steadily up until the early 1970s when NC machines were increasingly being controlled by computers giving true *Computer Numerical Control* (CNC). In a CNC system the stored program, holding the control data, can easily and quickly be modified providing a high order of machine versatility. No longer are machine tools dedicated to the production of a limited range of components. Computer Numerical Control systems, which usually have electronic feedback control, have many advantages over conventional Numerical Control systems; namely; greater reliability and flexibility of operation, provision of management information (i.e. machine status, downtime, cutting time). Tool positioning and cutter operating regimes can be tightly controlled in order to minimize tool wear, and to optimize machining. A CNC machine tool which can perform a range of activities is referred to as a *machining centre* (see Fig. 7.2).

CNC has, in the main, been applied to metal machining operations such as turning drilling and milling, partly due to the fact that non-

Fig. 7.2. Machining centre (courtesy Hepworth Engineering Limited).

machining applications such as fabrication and welding require a high degree of manual input. However, recent advances in CNC hardware and software has resulted in the use of computer numerical control in areas such as bending, cutting, grinding and punching.

Bending

CNC machines are now being used for precision bending of metal sheets. Metal cabinets, for example, can be accurately formed and some systems are also capable of punching holes in the metal as part of the same operation.

Cutting

Traditionally, sheet metal cutting has been performed using hand held gas cutting torches. This method was superceded, in some of the larger companies, by using optical profile cutters. Such machines relied on an optical sensor following a black line on a component drawing. Some machines employed a pantograph arm arrangement which meant that the drawing could be scaled up or down. However, the optical devices tend to be sensitive to marks or creases in the drawing. Other machines use a template copying system. Here a stylus follows the outline of the template and a flame cutter replicates the shape to produce the component from a blank sheet of metal. The one major drawback of this system is that a template has to be made for each new component and standard templates are often difficult or impossible to modify if minor profile changes are required. Depending upon the size of the component, these conventional cutting machines render the remainder of the blank sheet irrecoverable and hence a large amount of scrap is generated. Computer-aided manufacturing systems are now available which provide programs to generate graphically a number of different component shapes, which can be fitted in the most economical way onto a standard sized sheet of material.

Using the information from this *nesting* program, a flame cutter path program is generated to give optimum tool path efficiency and to minimize scrap.

Grinding

CNC grinding machines have been slower to develop than conventional CNC machines because of the complexity of the electronic control systems required. As the grinding wheel diameter is reduced during operation, by dressing and wear, the control has to continually calculate and update the wheel diameter to give the required positioning accuracy; typical diameter tolerance on a 50 mm diameter shaft would be ±2 μm. A CNC grinder is the only CNC machine that sharpens its own tool.

In precision cylindrical grinding, it is normally arranged that the longitudinal centre line of the work piece is coincident with the table top swivel axis, so that errors in taper are more easily adjusted.

7.4 COMPUTER-AIDED QUALITY CONTROL

Quality control which encompasses inspection, measurement and testing is a vital part of any manufacturing activity and is applied to ensure consistent high quality of manufactured goods.

A wide range of computer based instrumentation is being used for quality control. In particular, the increased sophistication of sensors and transducers is making it easier to carry out *pre-process, in-process* and *post-process* inspection and testing in an attempt to reduce machine time and to maximize resources.

Material stock can be automatically examined prior to machining to ensure that it can be accommodated by the machine. The workpiece can be monitored during machining to detect unfavourable cutting tendencies. Also, post-process inspection and test can check that a range of products is being manufactured to laid-down quality standards.

Inspection and Measurement

Effective inspection relies upon a number of crucial measurement parameters. The types of measurements usually taken include: linear measurement, ovality, concentricity, parallelism, and flatness and squareness.

Increasingly, these parameters are being checked by automatic devices which tend to reduce errors in measurement and increase the speed of measurement without detriment to accuracy.

CNC gauging and measuring equipment which uses electronic probes has been available for several years, but in-process gauging on the machine tool itself is a recent development. The use of a gauging probe as one of the tools in the maching centres tool post permits the machining centre to inspect its own work. Computer-controlled probes are used to check the dimensional accuracy of the workpiece.

A significant recent development is the use of adaptive controls to monitor the results of gauging and feed back the changes necessary to machine the component more accurately. Much inspection and measurement is still a post-process activity, and microprocessors are helping to reduce post-process checking and so improve manufacturing productivity.

There are a variety of ways of speeding up the measuring time. One method is to use several probes simultaneously, each of which inspects a different feature of the finished component. Data from the probes is processed by a microcomputer and the results displayed on a VDU

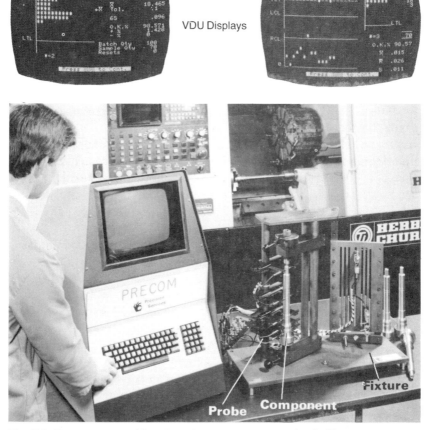

VDU Displays

Fig. 7.3. Post process, multi-probe, micro processor measuring facility (courtesy PRECOM).

screen. One system, as shown in Fig. 7.3, displays the dimensions on a screen as bars, showing the tolerance limits as a rapid check that the component is acceptable. Using this system, data can be analyzed for batches and statistical charts displayed to show machining trends. An alternative approach to post-process gauging is to use a three- or four-axis measuring machine where a single probe traverses the workpiece at high speed, approaching it from a variety of different angles and directions. The sequence of movements is programmed according to the requirements of the workpiece.

Gauging probe development now includes laser metrology, enabling very fine tolerances to be achieved consistently.

Random measurement can be carried out to build flexibility into quality control. This may be achieved, for example, by taking measurements with a hand held digital micrometer (eliminating the need for a jig), linked to a microcomputer.

Statistical quality analysis procedures (SQA) have been developed, in particular, for use in the manufacture of high-volume precision components, pistons, bearings, rings, etc. The objective is to achieve high quality assurance without the need for 100% inspection and test.

Vision systems are beginning to play a major role in inspection procedures. Vision systems are developing in conjunction with robots and range from simple photoelectric devices to more sophisticated pattern recognition systems where a component is recognized, the vision signal digitized, and compared with a stored pattern. Computer vision for inspection purposes has to meet a number of requirements. For example, the system must be capable of processing two-dimensional images. Also, it must be possible to extract useful information from the processed image, e.g. to carry out a dimensional analysis.

Testing

Automatic Test Equipment (ATE) has been in use for many years, but more recently is exploiting microprocessor capabilities. Automatic testing can be applied to many manufacturing industries but has found substantial use in the testing of electronic components and circuits (LSI, chips, printed circuit boards).

Many systems, for example, test every conceivable logical path through a PCB then transmit the results to the operator. Information can be obtained on whether the board has passed or failed, why the board has failed (e.g. a dry joint or faulty component), and indeed whether the test equipment is operating satisfactorily. Many thousands of connections and interconnections can be tested in a matter of seconds.

A major step forward in automatic testing is the use of robots to carry out remote tests in hazardous environments.

7.5   MANUFACTURING AIDS AND SYSTEMS

Robotics

Programmable robots began to appear in manufacturing industry in the late 1960s. Since then the growth of robots (particularly in Japan, Europe and the USA) has been phenomenal. Figure 7.4 shows the world robot population. By the late 1970s there was growing interest in robots and now all the leading manufacturing nations have their share of robot technology. Figure 7.5 gives the UK robot population.

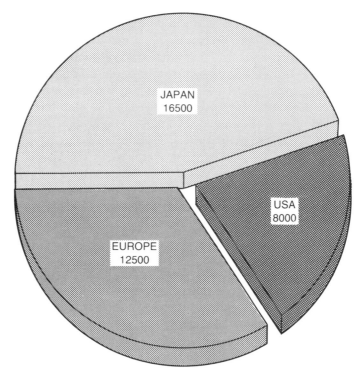

*Excluding Eastern Countries and the USSR.

Fig. 7.4. World robot population as at December 1983 (courtesy of British Robot Association).

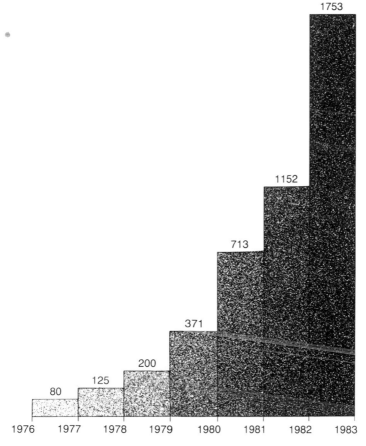

Fig. 7.5. UK robot population showing growth from 1976 (courtesy of British Robot Association).

Certain production environments are hazardous to human health, and the boring and repetitive nature of some tasks can lead a human operator to apathy and carelessness. Moreover, human beings become tired, ill, take holidays or go on strike. In contrast to this, industrial robots are resilient to hazardous environments and will carry out specified tasks repeatedly for long periods of time. It is clear why management should be investigating the possibility of industrial automation in which robots play a substantial part.

Currently manufacturers are relying more and more on robots and there is substantial evidence to show that industrial robots are highly cost effective. Robots can make the factory floor a safer and more productive place and the effective utilization of robots can be crucial to the survival of some manufacturing industries.

Robots can be used for many manufacturing tasks including: assembly, welding, paint spraying, machine feeding, testing and product packaging. Robots can handle a range of components (from the most delicate assembly routines to large fabrication jobs) in a variety of materials (glass, plastic, rubber, steel, etc.). Figure 7.6 shows the breakdown of UK robot manufacturing applications.

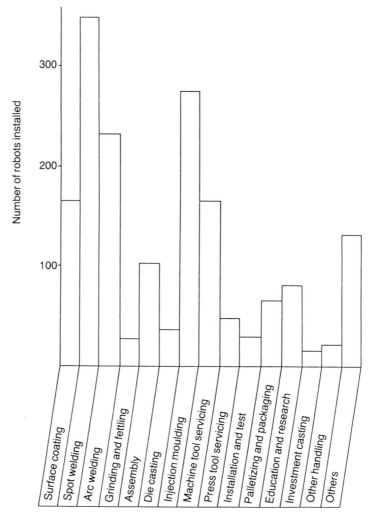

Fig. 7.6. Distribution of robots in manufacturing activities in UK as at December 1983 (figures courtesy of British Robot Association).

Automated Guided Vehicles

The position of machining centres and robot assembly and test areas are usually pre-determined when the factory floor layout is planned. The raw materials and finished products are traditionally taken to and from the working areas either by hand-cart, fork lift truck or conveyor belt.

The movement of work throughout the factory floor can be a time-consuming, labour-intensive activity and there are sometimes areas in a factory which are difficult or impractical to serve with a conveyor belt. Also, in the heavier industries the component parts are just too heavy for conventional conveyor systems to handle. To overcome these problems *automated guided vehicles* (AGVs) can be used for the transportation of parts in the factory.

AGVs are transporter cars (usually battery powered) which move loads from one specified location to another. AGVs are programmable and follow conductors which are either permanently embedded in the factory floor or alternatively a metallic strip with adhesive backing is laid on the floor. The latter allows greater route flexibility.

One system, the *MIKROTRUK* from Babcock Fata, as shown in Fig. 7.7, is capable of moving loads of up to 2000 kg to any desired

Fig. 7.7. Automatic Guided Vehicle (courtesy of Babcock Fata).

location on the shop floor. The MIKROTRUK has a variety of attachments such as flat top, elevating flat top, powered roller conveyer, telescopic chain conveyor and telescopic forks for handling a variety of components. A number of AGVs can be simultaneously in operation on the same factory floor without interfering with one another. Each truck has a high level of on-board intelligence with a capacity to store up to 8000 commands, and each is capable of handling quite long and complicated manoeuvre schedules without the need for any off-board processor.

Individual commands are entered in a logical, modular, high-level form, so that quite involved operations can be specified quickly and easily, and the modularity of the commands makes it possible to quickly change the role of a truck to keep it in line with the changing demands on the shop floor. Some AGV systems employ overhead infra-red monitors coupled into the manufacturing system software, making it possible to check the exact location and function of a truck at any given time. Most AGVs have flashing beacons, and wrap round touch-sensitive bumpers which immobilize the vehicle if it encounters an obstacle.

Flexible Manufacturing Systems

A *Flexible Manufacturing System* (FMS) is a production environment configured from machine tools to which a range of raw materials, parts or components is brought, loaded and removed by automatic means. This usually implies the use of conveyors, computer-controlled cranes, feeding robots and AGVs servicing the machine tools directly. A typical FMS layout is shown in Fig. 7.8. The key to the inherent flexibility lies in the ability to control all of the equipment through a computer system.

Machining programs and tool positioning parameters are automatically down-loaded from the computer to the machine tools according to the workpieces scheduled at each machine; the computer system also directs component handling conveyors or AGVs to transport part-finished work from one machine to its subsequent operation at another.

An FMS can therefore largely run unattended, and, although manual loading and unloading of pallets and fixtures may still be required, the machines can run virtually continuously, giving maximum utilization of plant.

The building blocks of any FMS system are the machine tool and a work handling mechanism (i.e. a robot). The work handling device should be capable of placing unmachined parts (rough castings, for example) on the machine tool and then removing the part after machining.

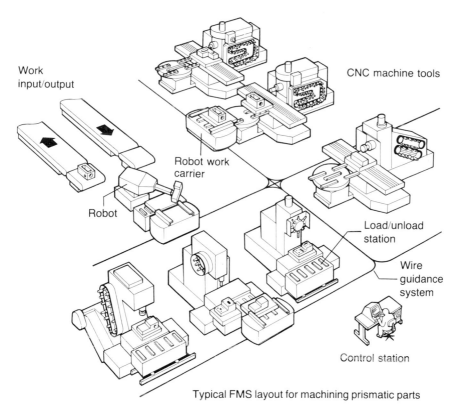

Typical FMS layout for machining prismatic parts

Fig. 7.8. Typical Flexible Manufacturing System (reproduced by kind permission of The Production Engineering Research Association).

There are two main philosophies employed for the relationship between the machine and robot. The first is the line approach, as in Fig. 7.9, where each machine tool has its own dedicated robot or alternatively the machine tools can be clustered and made to share a common robot, as in Fig. 7.10. This configuaration is often referred to as a *production cell*. Ideally, an FMS should use the best characteristics of each of these methods to arrive at a truly flexible system.

Whilst it is easy to control the individual elements of a system such as machining centres, robots, AGVs, conveyors, inspection equipment, and the automatic stores, the effective co-ordination of all these elements into a working FMS is the most difficult task for computer-aided manufacturing.

Every FMS is different and although there are standard routines becoming available for the control of the elements, it is the individuality

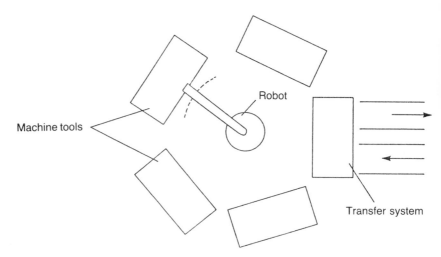

Fig. 7.9. Machine tool and robot in line arrangement.

Fig. 7.10. Machine tools clustered round robot to form a cell.

of each FMS which poses the problem. The reliability of the software is a major factor and is one which governs the introduction of FMS. In the early days of these systems the software was limited and inflexible. Today, however, good software is emerging for FMS installations in many industries, including applications in metal cutting, forming, wood-work, non-machining applications (such as component washing and packaging), electronic circuit manufacture, assembly and manufacturing processing.

The provision of a computer system for the basic control function allows other allied activities to be computerized — in particular: the scheduling of work to the different machines; the collection of statistical data on throughput and machine downtime; the control of tooling; the running of an automated stores area and the integration of a CAD/CAM system for part-program generation. Figure 7.11 shows how the

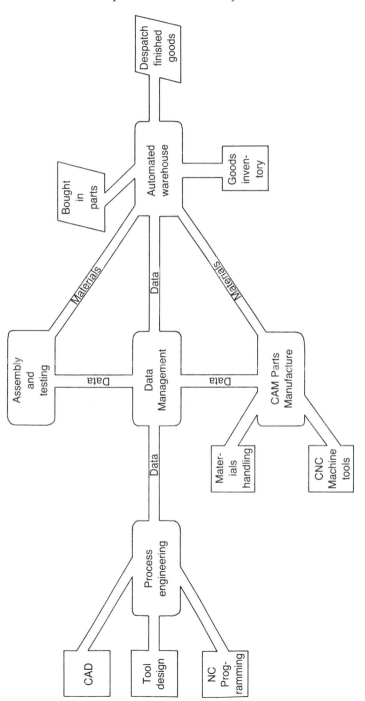

Fig. 7.11. Interdisciplinary activities of a flexible automated factory.

many design and manufacturing activities interrelate to give a flexible automated factory.

The computer system (software and hardware) for controlling FMS will vary from one system to another. One approach is to have a host computer, which may have a standby, directly linked to microprocessor-based controllers attached to each machine.

Typical FMS

An example of a working FMS is the British *SCAMP* system (600 group's computer-assisted manufacturing project). SCAMP manufactures a range of annular components such as gears, pulleys and shafts in a variety of materials. Nine machine tools perform turning, milling/drilling/tapping, grinding, gear cutting/shaping, hobbing and broaching operations, using robots to load the machines. A palletized conveyor system under computer control directs the workpieces at each stage to the relevant machining cells or loading sidings, using a bar-coded pallet identification system. Consequently, a batch of 50 components can be completed in less than 3 days, compared to 6–8 weeks using conventional manufacturing methods.

With the prospect of more powerful software being made available, the level of flexibility of programming and control of FMS (particularly in metal cutting processes) is increasing rapidly. As a result, estimates of the number of systems in use over the next ten years or so foresee a doubling or trebling on the present number.

There is a growing awareness, in manufacturing management, of the benefits of FMS. Most notable is the increasing interest among European manufacturers whose progress to date has been slower than their Japanese and American counterparts. Indeed as an example of this increase in interest, *SMT Machine Company* now arrange 4-day study visits to FMS sites in Sweden, for Manufacturing Managers and other interested parties.

Benefits of FMS

The benefits claimed for flexible manufacturing are manifold. The whole concept, however, is a compromise between high flexibility (general purpose equipment), and high productivity (dedicated automatic equipment).

Some of the major benefits include:

(i) Reduction in manpower due to the nature of automatic machining, machine loading and materials routing.

(ii) Integrated manufacturing of a wide variety of components, thus increasing maching utilization and throughput, without the need to re-configure or re-invest.

(iii) Accurate, up-to-the-minute, management information giving better control of batch loading and scheduling, and inventory control.

(iv) There is no requirement to store finished products as the response time of the environment means goods can be made to order (often called the *just in time* philosophy).

(v) An elimination of work-in-progress coupled with a reduction in both buffer stores and raw materials stock means that costs can be kept low.

(vi) The overall increased efficiency makes the company more competitive.

Advanced Manufacturing Technology (AMT)

AMT covers a range of equipment and techniques which form the basis for modern manufacturing systems ranging from sophisticated stand-alone CNC machines to fully integrated FMS, and high technology dedicated lines.

As well as CNC machine tools incorporating the latest refinements, in component loading, gauging, monitoring and control, AMT also covers component transportation control software and management information necessary to link these machines together as part of an Advanced Manufacturing System.

AMT installations are seen as ranging from stand-alone machines, usually of high sophistication, through linked cells and FMS to automated lines producing dedicated components. The relationship between AMT and FMS is shown in Fig. 7.12. All forms of AMT would require the following ingredients:

(i) Machine Tools
(ii) Tooling
(iii) Measuring
(iv) Monitoring
(v) Component loading, and
(vi) System control and Management

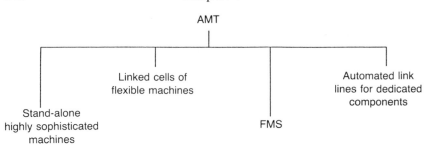

Fig. 7.12. Relationship between Advanced Manufacturing Technology and Flexible Manufacturing System.

The variety of components manufactured and the batch sizes required largely determine the form of AMT to use. Shown diagrammatically in Fig. 7.13.

7.6 COMPUTER-AIDED MANUFACTURING MANAGEMENT

Manufacturing industries rely on an efficient manufacturing (or production) system, including the effective deployment of manpower, plant,

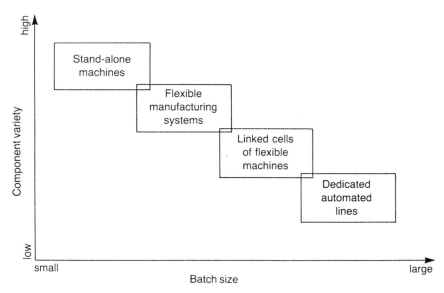

Fig. 7.13. The choice of various forms of AMT depending upon component variety and batch size (reproduced by kind permission of Traub Limited).

supplies, and capital. Traditionally, manufacturing management has relied on human expertise, but the sophistication of Computer-Aided Production Management techniques (CAPM) means that managers in industry are rapidly computerizing their manual systems, particularly in the following areas:

Process and capacity planning
Production control
Inventory control
Materials requirements planning

Figure 7.14 illustrates some of the more important aspects of computer-aided production management.

Companies which have recognized the need for CAPM, and have introduced or developed systems to provide it, enjoy many advantages over their competitors. For example, they can set competitive delivery dates quickly and with confidence, minimize stocks and work-in-prog-

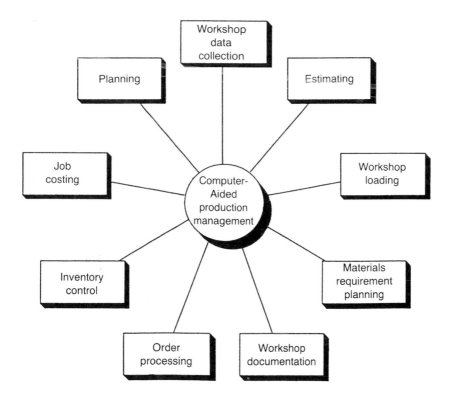

Fig. 7.14. Main aspects of Computer-Aided Production Management.

ress, utilize plant more efficiently and reduce manufacturing lead-times.

The main aspects and features of computer-aided production management are now given.

Computer-Aided Process Planning

Process planning is the establishment of an outline of steps required to produce a finished component. This plan covers a number of functions in most companies, the main ones being: component routing, method description, time generation, standards creation and maintenance, and NC programming. Each of these involves the preparation of documentation that is used for the instruction of the personnel involved in manufacture or instructions for the components of an FMS. Computers can be used to aid the preparation of this documentation, enabling the planning staff to assess the optimum method of production, giving speedy access to machine tool types and characteristics. The computer can also provide direct links to a CAD system in terms of geometric definition and component attributes. It also has links to the control aspects of manufacture.

The Need for Computers

There are many variables involved in the production of components — for example, components have varying degrees of complexity, materials used are different, quality and tolerance standards (although generally demanding) can differ, many kinds of machines are used, a variety of operations and a whole host of different workers may be involved in the production of the finished component. Without a set plan to follow, production would be almost impossible.

A computer is used to examine and co-ordinate all of these variables in order to achieve an efficient plan with the following benefits:

   (i) Reduction in clerical effort.
  (ii) Improved consistency of applications of data.
 (iii) Improved data accuracy.
 (iv) Standardized manufacturing methods may be used.
  (v) Records can be kept up to date.
 (vi) Logic can be captured and used by anyone.

A CAPP system usually consists of five main elements as follows:

 (1) Coding and classification.
 (2) Database creation and maintenance.

(3) Logic processor.
(4) Documentation production.
(5) File maintenance.

(1) *Coding and Classification*

The application of computers in process planning forces companies to be logical and to think about standardizing their approach to manufacture. This is because the computer has to act upon certain logical rules. It is necessary, therefore, to have some means of identification of individual parts, similar parts, individual routings, and similar routings within the numbering system that identifies these features. Many companies have their own in-house numbering systems. Some are sequential, some alphanumeric, but most suffer from the problem that when they were conceived they were not structured for use within a computer.

Recently, many new coding and classification systems have been designed specifically for use on the computer. The better known of these are *MISCLASS* and *CODE*.

Both systems try to identify similarities between geometry and manufacturing methods rather than identifying uniqueness. This tries to ensure that people search for similar parts and methods of manufacture before designing or planning a new component.

Following the introduction of a classification and coding system, it is possible not only to standardize on routings, geometry etc., but also to maintain component history and ensure that changes are transferred to other similar components if required.

(2) *Database Creation and Maintenance*

The likely contents of a database are as follows:

Synthetic Data — This data covers the small elements of time that the company uses in individual manufacturing steps.

Work Patterns — Within the database should be held the normal association of elemental data that are used in building up the system.

Documentation Text — This is useful in that it ensures that planners use the same text in relation to a job and the same descriptions of parts, so that different people do not describe the same things in different ways.

Tabular Data — A major part of the database is the record of manufacturing facilities available. This can be in terms of capacities, capabilities, tooling, gauging, etc. — most of the data being held in tabular form.

(3) *Logic Processors*

The logic processor is the heart of the CAPP system and ensures that different people plan the same manufacturing method in the same way. Essentially, there are two types of logic processor — a low-level and a high-level processor. The former holds, within the computer, the detailed method of manufacture at operation level, whereas the latter is capable of being specified because the product being manufactured is fairly standard or falls into some logical grouping.

(4) Documentation Production

As the manufacturing process is determined by the CAPP system it is also possible to produce 'shop floor' documents for controlling manufacture.

   Documents typical of a CAPP system include method sheets (including tooling details, speeds, feeds, etc.), routing sheets (giving details of work centres, where the job passes) and tool fitting sheets (details of types of tool required prior to manufacture).

(5) *File Maintenance*

This covers the storage and control of records, and the ability to retrieve and edit data as required. Normally within a CAPP system there is a need to store output information in one form or another. New, deleted or amended records must be identified as such, so it is important that the database management system is well organized, with no duplicate or obsolete records kept.

   There are a number of CAPP systems on the market which can be used for a specific process planning need. These include *CAPES, CAPP, LOCAM, MICROCAPE* and *SOFIE.*

Computer-Aided Production Control
Computer-Aided Production Control (CAPC) is a modern term for a group of techniques which have been evolving since the 1960s. The basic principles were established long before that, along with the recognition of production control as a discrete management task. The main aim of an effective production control system is to try to ensure the manufacture of suitable products as cheaply as possible whilst maintaining acceptable quality standards.

   Production control is a broad-based concept with many separate but interconnected entities — such as data capture, network planning, estimating, information handling, materials requirement planning and inventory control.

Network Planning

This is a method of production planning by decribing the project as a *network* and then carrying out an analysis on the network.

The analysis involves:

(1) Constructing a network diagram to represent the project to be undertaken, indicating the sequence and interdependence of all necessary jobs in the project.

(2) Determining the duration of each of the jobs.

(3) Performing network calculations to determine the overall duration of the project and the criticality of the various jobs.

(4) Re-assessing the network and modifying job durations in order to complete the project within the required time.

Network analysis techniques have been used as a manufacturing tool for many years. Critical path methods are well established and the fact that a network can be described numerically lends itself to computerization.

Today there are a number of sophisticated packages available such as *CPM* (Critical Path Method) and *PERT* (Programme Evaluation and Review Technique), for effective network planning.

Estimating

It is difficult to satisfactorily estimate time schedules for manufacturing processes due to unforeseen factors such as machine downtime, tool breakages, operator illness, etc. The conventional manual estimating task can be made more accurate by using computing methods. It has been found that in excess of 30% improvements in output are obtainable when computerized estimating is used.

Inventory Control

Inventory control is an important element in a production control system. If stock levels are too low, production may be delayed or stopped; if too high, a large amount of capital would be tied up. Computer-aided inventory control can be used to optimize the control of stock.

Early software packages for inventory control were based on simple accounting routines, as it was realized that the techniques used for handling monetary transactions could easily be applied to the control of physical stock.

Inventory control packages developed rapidly over the years, with features such as multiple locations being added. It is also common to include statistical inventory control techniques so that stock levels could trigger the re-ordering of components automatically. The more sophisticated versions link to the sales and purchase order files to give overall control of stock movement.

A computerized inventory control system would handle the following:

(i)   Issues against sales order and returns.
(ii) Recording and cancelling purchase orders.
(iii) Issuing of individual parts kits.
(iv) Stock adjustments.

The system would normally indicate the stores location (i.e. rack and bin) of a given part and may also issue early warnings of stocks running at a low level. Most systems have a parallel costing function so that the value of items in stock and in progress on the shop floor is readily available to management.

Material Requirement Planning

Many production control packages were developed throughout the late 1960s — such as shop loading and capacity planning software. However, the largest step forward involved the development and use of Material Requirement Planning (MRP). By the late 1970s many companies were using MRP to schedule manufacturing operations and since that time MRP has come to be regarded as an essential element in any comprehensive production control system. As a result, a growing range of MRP software is being made available commercially.

Materials requirement planning is a technique for managing production inventories by taking into account the specific timing of material requirements. The aim, then, is to minimize the investment in this area of inventory without impairing a given production plan.

One of the first industries to adopt MRP on a large scale was the motor industry, particularly in the USA. Most of the larger companies had built up files defining the content of their products for costing purposes. These product structure files, as they were called, could then be applied to determine what material to order, based on the projected product production over a forthcoming period. Thus the demand for material is dependent on the production programme.

MRP broke down the overall demand through the product structures by offsetting the dependent demand against sub-assembly and component stocks on hand. It could then place orders for new material while taking account of the lead time required to obtain it. Any independent requirement for spares could be added to the net demand at the end of the process.

MRP, when properly applied, provides substantial savings in inventory costs and has the added benefits of reducing obsolescent stock and, improving material flow within storage areas.

There are a large number of MRP software packages on the market, including: *PICS* from *IBM*, *COPICS* and *MASE* from *Martin Marietta Corporation*.

Chapter 8

Machine Tool Control

8.1 INTRODUCTION

Ever since machine tools have been used as a means of quantity production in manufacturing industry, there has been a trend to invent and develop devices which make the machines *automatic*. These innovations were in order to reduce the manual content of machining as labour costs have increasingly become the largest single cost element in the total cost of a product. The workhorses of a manufacturing industry are its machine tools. Conveyors and robots play a part in transferring and loading materials and components, but it is the machine tools which turn, mill and grind the components that will eventually find their way into the market place. These days, it is probably easier to pick out the processes which have not been influenced by automatic machine tool control rather than those which have. Numerical Control (NC) has found its way into processes such as grinding, honing, thermal profiling, injection moulding, shot peening and electron-beam welding, and with its role in robotics, NC looks set to be the *major* element in a production system.

This chapter explains the concepts involved in Numerical Control (NC) and Computer Numerical Control (CNC). The elements of a CNC system are outlined and recent enhancements to CNC machines are discussed. The methods of programming CNC machine tools are given and a simple programming case study shows how a program is constructed for a particular machine. The more advanced control concepts such as Direct Numerical Control (DNC) are explained and finally the way in which a CNC machine can be linked with the CAD function is described.

8.2 NUMERICAL CONTROL CONCEPTS

Machine tool control systems usually fall into one of two categories: *open-loop* control or *closed-loop* control. In an open-loop control system a command is directed to the machine to carry out a given operation. The tool will move to its ordered position, carry out the operation, then stop. As the loop is open there is no feedback, the control system

156

generates no return signal and there is no indication that the tool has, in fact, moved to its desired position.

In a closed-loop system, feedback is built into the control. As the tool moves to a new position, its movements are monitored automatically. A signal is returned to the control unit, and if the tool is not in the correct position it is automatically adjusted until it is. Most modern numerically controlled machine tools employ a closed-loop control system.

Control System

The program from which the command signals are derived is stored in numeric form. The numbers can be encoded on tape and the tape can be run many times, through a decoding device attached to the machine. The fact that the numeric data is permanently encoded leads to consistency and the elimination of errors.

To understand the basics of a numerically controlled machine tool, consider the diagram in Fig. 8.1.

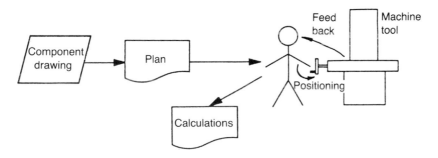

Fig. 8.1. Conventional manual closed loop control system.

With a conventional system, some pre-planning takes place before the drawing and raw materials are passed to the operator. The operator interprets the drawing, carries out a few calculations, loads the workpiece and then turns the machine handwheels to make the component. The operator monitors the tool position and makes any adjustments to ensure the component is made to within pre-defined tolerance limits. This is then a closed loop system.

The human operating functions, i.e. turning handwheels, controlling coolant flow, etc., may be replaced by numerically controlled operating functions as in Fig. 8.2.

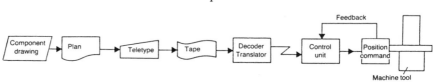

Fig. 8.2. Numerically controlled operating functions.

The machine tool is equipped with a control until which feeds positional command signals to servo motors which control the tool post position, in the case of a lathe, or the machine bed, in the case of a vertical milling machine. A feedback signal is fed to the control unit from a position monitor to control the position of the tool or bed.

The monitoring systems used for numerically controlled machine tools are, in effect, position transducers which enable feed-back signals relating to the bed or tool position to be transmitted to the control unit. Figure 8.3 shows, diagrammatically, the layout of a monitoring system. Various monitoring systems are in use but typical ones are (a) analogue devices such as potentiometers or synchros, and (b) digital devices such as optical gratings and digital transducers.

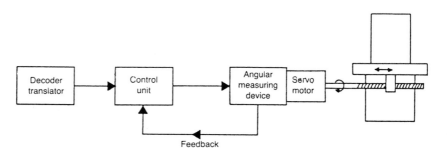

Fig. 8.3. Diagrammatic representation of NC monitoring system.

In a non-operator environment the component drawing must be translated into a series of co-ordinate information and command signals that the machine can understand. This series of commands is called a *part-program.*

The majority of program codes used for NC machines are based on paper tape codes. A selection of codes is given in Fig. 8.4. The part program, once read by the decoder, becomes a set of instructions that the machine can understand.

In most NC machining systems, tapes are used for transmitting information given on the drawing to the decoder and again used for

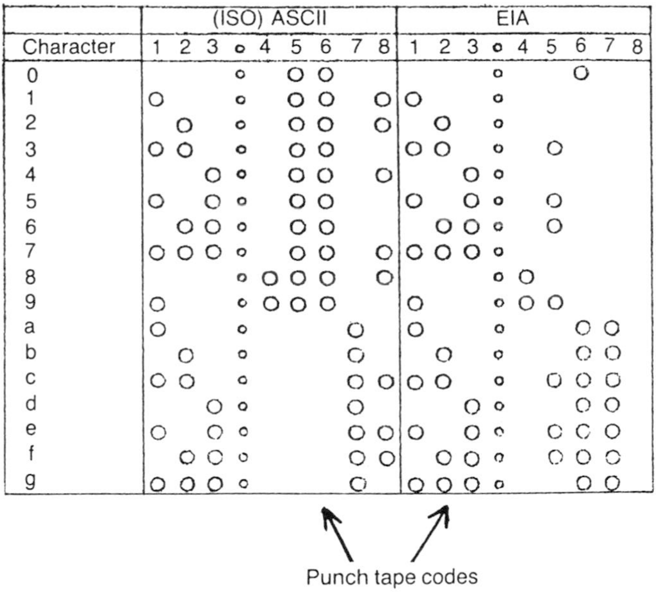

Character	(ISO) ASCII									EIA								
	1	2	3	o	4	5	6	7	8	1	2	3	o	4	5	6	7	8
0				o		O	O						o			O		
1	O			o		O	O		O	O			o					
2		O		o		O	O		O		O		o					
3	O	O		o		O	O			O	O		o		O			
4			O	o		O	O		O			O	o					
5	O		O	o		O	O			O		O	o		O			
6		O	O	o		O	O				O	O	o		O			
7	O	O	O	o		O	O		O	O	O	O	o					
8				o	O	O	O		O				o	O				
9	O			o	O	O	O			O			o	O	O			
a	O			o			O	O	O	O			o			O	O	
b		O		o			O	O	O		O		o			O	O	
c	O	O		o			O	O		O	O		o		O	O	O	
d			O	o			O	O	O			O	o			O	O	
e	O		O	o			O	O		O		O	o		O	O	O	
f		O	O	o			O	O			O	O	o		O	O	O	
g	O	O	O	o			O	O	O	O	O	O	o			O	O	

Punch tape codes

Fig. 8.4. A selection of codes used in numerically controlled programming.

transmission of data from decoder/translator to the control unit.

For generation of the part-program in the simplest NC system, the operator usually types the co-ordinate and control information into a teletype which then produces a tape, usually in paper. It is possible (and usual in shop floor environments) to copy the information on to a more robust media such as *mylar*. Magnetic tapes have been used to transmit control data from the decoder to the machine controller but are being replaced by hard-wired systems.

Development of NC

NC machines were initially designed and brought into manufacturing because of their ability to produce highly complex parts consistently. This gave improved quality and a reduction in scrap, but required accurate machines with reliable control systems. For the machines to be controllable in a consistent manner, their whole design required improving and this led to new designs of machines which were not only controllable to fine limits, but were also more powerful in terms of material removal rates and cutting speeds.

As the quality of designs improved still further, designers found it possible to widen the capability of machines so that drilling, milling and boring could be carried out on the same machine. The same evolution happened with lathes and turning machines.

Different operations require different tools and, to speed up the tool change process, automatic tool changes were introduced on vertical machines and additional rotating turrets on lathes.

8.3   COMPUTER NUMERICAL CONTROL

As computer technology improved over the years, NC underwent one of the most rapid changes in the history of production engineering. Integrated circuits were used in NC controllers. Computer hardware became less expensive and more reliable and NC control manufacturers introduced *read only memory* (ROM) technology. ROM was typically used for program storage in special purpose applications, leading to the appearance of the Computer Numerical Control (CNC) system. CNC was successfully introduced into practically every aspect of manufacture. The coming of CNC brought some substantial improvements to the operation of NC machines. In particular, it enabled the paper tape decoder unit to be bypassed and allowed programs to be entered and edited at the machine. It eased the maintenance and repair of controllers in a number of ways and it opened up the possibility of linking machines together, through their computers, to form part of a larger system. It was made possible to read the machining instructions into memory and operate the machine direct from the memory.

Elements of a CNC Machine Tool

For simplicity, CNC machine tools will be considered to fall into one of two categories. *Lathes* (turning machines or horizontal machining centres) and *Milling machines* (vertical machining centres). The basic anatomy of these two types of machine will be discussed and then the elements of a CNC system, common to both types, will be explained.

Lathe Anatomy

Lathes are used for the manufacture of rotationally symmetric components by turning, boring, drilling, or thread cutting from solid bars, hollow tubes, fabricated rings or drum-like castings.

A typical lathe is shown in Fig. 8.5. The main components of the lathe are:

(i) *The Chuck* — The chuck is used as a method of holding the

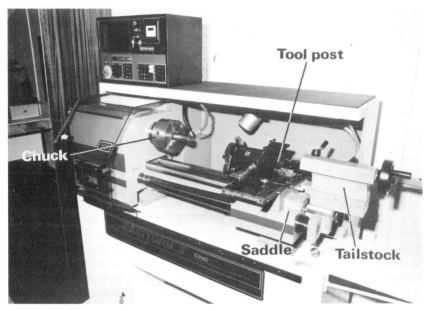

Fig. 8.5. The main components of a turning machine (courtesy Denford Machine Tools Limited).

workpiece and rotates at speed to facilitate metal cutting. The chuck is driven from a motor via gears in the headstock. Spindle operating speeds vary from 50 rev/min to about 5000 rev/min, depending upon the material being cut.

(ii) *Tailstock* — This is used to give additional support to long work pieces. It prevents movement of the workpiece about the machining axis.

(iii) *Saddle* — The saddle is the part of the lathe which carries the cross slide. The saddle traverses the lathe bed from the headstock to the tailstock providing movement in the z direction. Movement at 90° to the lathe bed is facilitated by the cross slide. This is mounted on the saddle and carries the tool posts. Hence, movement of the tool is possible in the x and z axes; therefore lathes and turning machines are called 2-axis machine tools.

(iv) *Tool Posts* — The tool post is situated on top of the cross slide and is used to hold the cutting tools. There can be two tool posts on the cross slide, a front tool post and a rear tool post, as shown in Fig. 8.6.

Plan view

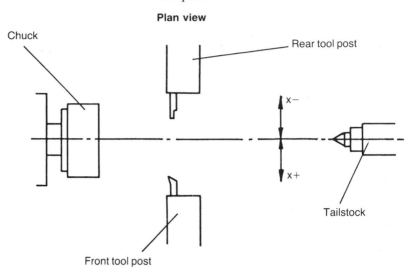

Fig. 8.6. Position of front and rear tool posts in relation to the chuck face.

(v) *Tool Turret* — The turret is used in addition to the tool posts for holding cutting tools. Turrets have a number of stations providing a variety of tools, and are capable of being indexed automatically. A turret may be mounted vertically or horizontally, depending upon the type of machining centre. Figure 8.7 shows a twelve-station turret with twelve tools mounted, six for internal diameters and six for external diameters.

(vi) *Ancillaries* — To ensure safe efficient operation of the lathe, certain ancillary features are usually incorporated. These include: coolant jets, providing a continuous flow of cooling medium to the cutting tool/workpiece interface; machine lighting to aid accurate tool settings; and a guard to prevent swarf and chippings leaving the work area.

Milling Machine Anatomy

Milling machines are used for the machining of components which have irregular shapes or require operations which cannot be provided by a turning machine. Typical milled components include dies, plates, brackets, or components with slots or holes in them.

A typical milling machine is shown in Fig. 8.8.

The main components of the milling machine are:

Fig. 8.7. A twelve-station turret with six tools for internal diameter and six tools for external diameter (reproduced by kind permission of Hepworth Engineering Limited).

(i) *Table* — The table or bed is used as a mounting surface for the workpiece. The workpiece can be secured to the table by clamps or can be magnetically attached. The table has mutually independent movements in three directions x, y and z, shown in Fig. 8.8. The table movement provides the three basic axes for milling operations. On some machines the table may also rotate.

(ii) *Machine Head* — The machine head, attached to the body of the machine, is normally in a vertical plane, although for some milling operations the head can rotate about a horizontal axis (angle θ in Fig. 8.8) to a fixed position. The head can be removed and replaced by an arbor for the attachment of circular cutting

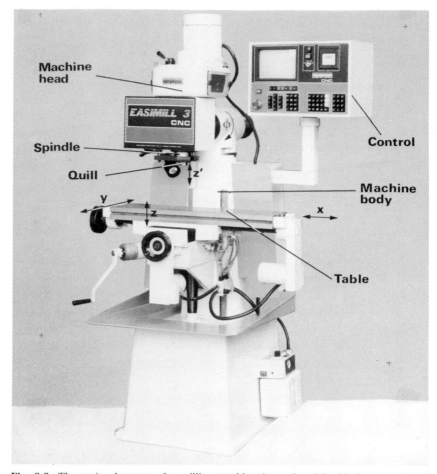

Fig. 8.8. The main elements of a milling machine (reproduced by kind permission of Denford Machine Tools Limited).

tools or a horizontally-mounted indexing tool turret.

(iii) *Spindle* — The spindle is used to hold the cutting tool which may be a milling cutter, drill, reamer or tap. The spindle rotates at speeds of up to 3000 rev/min to facilitate metal cutting. Spindles are usually belt driven from a variable speed motor. On the larger milling machines there is the facility for a selection of tools. These tools are stored in a carousel and can be changed automatically. Figure 7.2 shows a milling machining centre with a selection of tools in a carousel.

The cutting tool is allowed to traverse by a limited amount in a vertical plane (s') by movement of a *quill*. This quill is essentially a sleeve which is located in the spindle body, and is useful for repeat cutting operations without the need to move the table in the vertical plane.

(iv) *Ancillaries* — As with the lathe, to ensure safe, efficient milling operations, milling machines usually incorporate additional features. These include coolant jets (spray mist coolant often used), splash guards and machine lighting.

Common Machine Elements

Broadly speaking, there are a number of machine elements which are common to all CNC machine tools. A brief description of the main ones is given below:

Servo Motors and Drives

The pulsed electrical output from the controller is used to drive *servo motors* to position the work table or tool post to the desired position for machining. The drive system for the table or slides is called a *servo drive*. The servo motors which drive these components have to be very accurate. Two basic types of motor exist.

(a) *High Frequency Electric Motor*

These motors are operated on a 3-phase supply at 380/440 volts. The one drawback with this type of motor is that maximum power is limited and may be a disadvantage when heavy metal removal rates are required.

(b) *Hydraulic Motor*

A hydraulic servo motor consists of a motor and rotary pump. Fluid is pumped into the motor under pressure causing the rotor to turn and hence drive the machine lead screws.

Motors are used in the range 2 to 11 kW giving an accuracy of response of 0.005 mm. The hydraulic fluid is metered to the motor rotor by means of an electro-hydraulic transducer which receives its command signal from the controller.

CNC Positioning Methods

The way in which the tool is positioned during machining depends upon the positional control system used. There are a number of different philosophies on how this positioning should be achieved but most systems are based on one of two methods.

(i) *Point-to-Point* — The main function of the point-to-point positioning control system is to position the tool from one point to another within a co-ordinate system. The positioning may be linear in a plane or linear and rotary. Each tool axis is controlled independently and thus the programmed motions may be simultaneous or sequential. The important factor pertaining to point-to-point systems is that machining can only take place once positioning is completed. Point-to-point control systems are typically used for applications such as boring, tapping, punching, riveting and drilling.

(ii) *Continuous Path* — This type of control system is employed on the more sophisticated CNC machines. It is more versatile than point-to-point but its operation is more complex. The system generates a continuously controlled tool path by interpolating intermediate co-ordinates, which means the system has the capability to compute the points on the tool path.

Most continuous path systems also have circular or parabolic interpolation features and possess programming capability to handle as many as five independent axes.

Controller

Modern controllers are microprocessor-based logic devices. The controller is the link between the user's program of co-ordinates and instructions and the machine tool and work motion. The controller performs many functions as it converts numerical control instructions input by the programmer into program controlled instructions for the machine tool. The main functions of a controller include:

1. Reads, checks and interprets the part-program.
2. Transmits error and warning messages to the operator.

3. Activates machine ancillaries, e.g. lighting, coolant, etc.
4. Starts the main drive motors and selects spindle or chuck speed.
5. Ensures that the correct tool is selected.
6. Sends signal to servo motors to drive tool and workpiece to required position for machining.
7. Accepts feedback signals from position monitors and keeps a check on cutter location.
8. Calculates tool offsets and tool wear compensations.
9. Sends signals to automatic tool changers.
10. Stops the machine after operation and informs operator of completion; alternatively shuts down the machine if problems arise.

Memory

EPROM — non-volatile memory — is used to store control-programs in CNC controllers.

Bubble memories are being used for some CNC machines and tests have shown that this type of memory is less sensitive to workshop environments (vibration, temperature, dust particles) than EPROM.

RAM memory is used to store part-programs. Capacities vary but typical memories will hold about 8000 characters of program instructions.

Programs can be entered into memory in a variety of ways, either manually via buttons on the control panel (referred to as manual data input), by magnetic tape or paper tape via a tape reader, or data can be down-loaded directly into memory from magnetic disk or other remote computer equipment.

VDU and Control Panel

The VDU (Visual Display Unit) and control panel provide a link between the operator and the machine tool. A typical control panel with integral VDU is shown in Fig. 8.9.

VDU

Typically, visual displays used in CNC equipment are 230 mm green CRTs with 16 lines and 32 characters per line.

The VDU can be used to display the following:

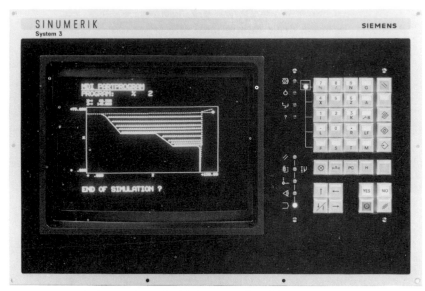

Fig. 8.9. A typical control panel with integral VDU (photograph courtesy of Siemens Control and Automation Systems).

Memory display — Sections of memory can be displayed for verification of such items as sub-routine numbers, spindle speed values, feed rates and so on.

Data input line — As data is being input, either manually or direct, the current program line and a few previous lines are displayed.

Positional Readout — Actual real-time values of all axes are displayed as the machining operation progresses.

Status Signals — Indicators for important modes and states of operation can be viewed.

Part Program Simulation — To enable tool path visualization.

Control Panel

The control panel consists of a series of buttons and switches, as shown in Fig. 8.9, and is for use by the operator to communicate with the controller. The operator will use the control panel for the following:

Initiation — This involves the manual setting of the machining datums (origins) to which all other dimensions relate. Tool offsets are also set manually.

(a) Absolute

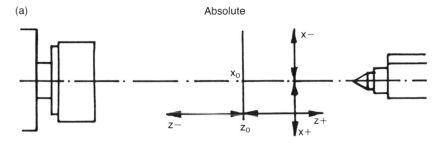

(b) Incremental

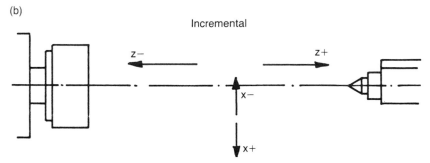

Fig. 8.10. Machine axis format.
    (a) Absolute.
    (b) Incremental.

Program Editing — It is possible to edit a program once stored in memory by calling up the required line(s) and re-entering the data at the control panel.

Manual Override — The operator can override the automatic control and use the machine tool as a conventional machine. This override feature is useful when establishing new machining regimes or unconventional tooling arrangements.

8.4 PROGRAMMING CNC MACHINES

A part-program for a CNC machine is made up of a series of lines of instructions (*or BLOCKS*). Each block has a number and is followed by a code. The code may also be followed by co-ordinates or other details such as feed rates.

Types of Code

G Code — The main code used in CNC programming is called the *preparatory function* code or 'G' code. For example G01 indicates movement of the tool/workpiece in a straight line and G02 means circular motion in a clockwise direction.

G codes vary slightly from machine to machine but the more common machining activities have standard codes. Table 8.1 shows typical preparatory function codes for a turning machine.

An example of a program block incorporating a G code is shown below.

N04	F120	X7	G01
Block No. 4	Feed rate in mm/min	7 mm from *x* origin	Circular interpolation

This instruction is for a turning machine and it instructs the controller to traverse the tool in a straight line to a position 7 mm from the pre-defined *x* origin, keeping the remaining axes constant. The tool feed rate is 120 mm/min.

M Code — Another type of code used in prgramming is the *miscellaneous code* or 'M' code. M codes are used to indicate the start and stop of a program or to start the machine spindle rotating in a given direction. Again, M codes vary for different machines — the standard M codes are given in Table 8.2.

An example of a program block incorporating an M code is given below.

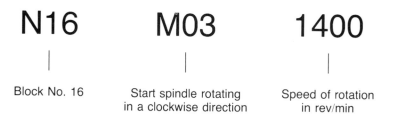

N16	M03	1400
Block No. 16	Start spindle rotating in a clockwise direction	Speed of rotation in rev/min

This block instructs the controller to start the spindle rotating at 1400 rev/min in a clockwise direction. Other codes exist, e.g. B codes which cover such items as bar loading magazine, rotating tool drive, spindle positioning and measuring equipment parameters.

Table 8.1

PREPARATORY FUNCTIONS (G-CODES) IN TURNING

Standard G-code	*Function*
G00	Positioning (rapid traverse)
G01	Linear interpolation (cutting feed)
G02	Circular interpolation (clockwise)
G03	Circular interpolation (counterclockwise)
G04	Dwell cycle
G10	Offset value setting by program
G20	Inch data input (G70 on some systems)
G21	Metric data input (G71 on some systems)
G22	Safety zone programming
G23	Programmed crossing through safety zone
G27	Reference point return check
G28	Return to reference point
G29	Return from reference point
G30	Return to 2nd reference point
G31	Skip cutting
G32	Thread cutting
G34	Variable lead thread cutting
G36	Automatic tool compensation
G37	Automatic tool compensation
G40	Tool nose radius compensation cancel
G41	Tool nose radius compensation left
G42	Tool nose radius compensation right
G50	Programming of absolute zero point
	Maximum spindle speed setting
G65	User macro call command
G66	Modal user macro call command
G67	Modal user macro call cancellation
G68	Mirror image for double turrets ON
G69	Mirror image for double turrets OFF
G70	Finishing cycle
G71	Stock removal in turning
G72	Stock removal in facing
G73	Pattern repeating
G74	Peck-drilling in Z axis
G75	Grooving in X axis
G76	Thread cutting cycle
G90	Cutting cycle A
G92	Thread cutting cycle
G94	Cutting cycle B
G96	Constant surface speed control on
G97	Constant surface speed control cancel
G98	Feed per minute
G99	Feed per revolution

Table 8.2

MISCELLANEOUS FUNCTIONS (M-CODES)	
M-Code	*Function*
M 00	Program stop
M 01	Optional stop
M 02	End of program and tape rewind
M 03	Spindle start CW
M 04	Spindle start CCW
M 05	Spindle stop
M 06	Tool change
M 08	Coolant ON
M 09	Coolant OFF
M 19	Spindle orient and stop
M 21	Mirror image X
M 22	Mirror image Y
M 23	Mirror image OFF
M 30	End of program and memory rewind
M 41	Low range
M 42	High range
M 48	Override cancel OFF
M 49	Override cancel ON
M 98	Go to subroutine
M 99	Return from subroutine

Note: Numerous other M-codes are available on various machines, as indicated by the respective manufacturer.

Machine Axis Formats

The machine axis format depends upon the machine tool being considered. Generally speaking, however, turning machines have two axes and vertical milling machines have 2½, 3 or more axes. (2½ axis implies that the tool and work table cannot move simultaneously, thereby preventing the generation of true 3-D contours.) The position of the tool (in the case of a lathe), or the work table (in the case of a milling machine) can be defined in terms of *absolute* units relative to a pre-set origin or in *incremental* units relative to the previous tool/workpiece position.

Turning Machines

X and Z axis notation is used for turning machines. The Z axis lies along a line which runs through the centre of the chuck (i.e. the axis of rotation). The X axis is at 90° to the Z axis. The absolute axis format is shown in Fig. 8.10a. The datum, X_o, for a turning machine, is always on the centre line although the datum Z_o can be at the end of the workpiece or, alternatively, the face of the chuck may be used as the Z_o datum. The incremental movements are taken relative to the previous tool position, as shown in Fig. 8.10b.

Milling Machines

The basic axis format for a vertical milling machine is shown in Fig. 8.11a where the X and Y axes are at 90° to each other in the same horizontal plane and the Z axis is perpendicular to the plane. The origins can be pre-defined by the operator although it is usual to define X_o and Y_o at a position away from the workpiece, as shown in Fig. 8.11b. The X_o, Y_o origin is commonly termed the *home position*. The home position, or machine origin, is physically set by the machine manufacturer and is used to synchronize the machine with the control and to establish a start point for measuring the length of travel in the X and Y axes. In fact, some machines cannot be started up until this referencing has been carried out. Referencing is achieved by returning the table to its home position and initializing the control so that the displayed X and Y values read zero. Z_o is often defined at some convenient point on the workpiece, usually the top surface on a flat component, as illustrated in Fig. 8.11c. The Z axis may also have its own home position. As with turning machines, milling co-ordinates may be expressed in terms of absolute or incremental units.

Tool Offsets

In addition to the home position or machine origin described earlier, there may be additional potential origins connected with a CNC machine, such as jig or fixture origin, workpiece origin, or program origin.

All these origins may be different or they may be interrelated in some way. The word *offset* has developed over the years from fixed origin, fixed zero, to zero offset. For reasons of simplicity, only tool offset relative to the workpiece will be considered here. It is possible with positional monitoring systems to ascertain the position of the spindle quill on a milling machine or the tool post on a turning machine, but because cutting tools have differing lengths and varying thickness, a method is required to establish that different tools always have their cutting edges in the correct position relative to the workpiece. The relationship between the cutting edge of the tool and a known position on the machine is called the tool offset and is best illustrated by way of simple examples.

(a)

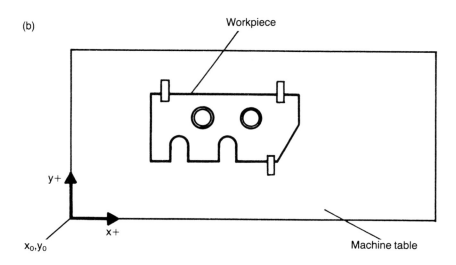

(b)

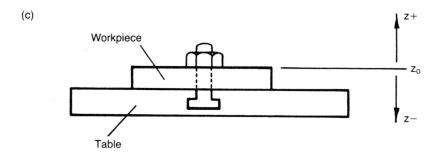

(c)

Fig. 8.11. (a) Basic axis format for a vertical milling machine.
(b) Milling machine table origin.
(c) Milling machine vertical plane origin.

(i) *Tool Length Offset for Milling Machine*

Consider the diagram shown in Fig. 8.12a. For drilling purposes the tip of the drill is positioned on top of the workpiece. With the machine table in a fixed Z plane, the lower end of the quill protrudes below the spindle face by an amount x. The position of the quill relative to the spindle is entered by the operator as the Z_o position. The physical position of the quill is known by the controller which equates the physical location with the operators Z_o and the tool length offset is established for this tool. If, however, a second operation is required using a different tool type, the offset will have to be re-established for the new tool. Figure 8.12b shows the same machine configuration with a different tool in the quill. The new tool is shorter in length than the previous tool, so, for the tool tip to be on Z_o axis, the quill has to protrude out of the spindle by an additional amount δx, giving a total protrusion $x+\delta x$. A new offset will now have to be calculated by the controller.

Preparatory function codes can be used for tool length compensation, allowing the operator to perform drilling, boring,

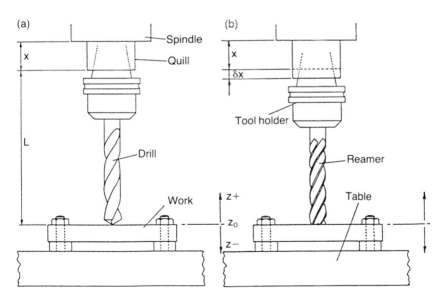

Fig. 8.12. Tool length offset for milling machine.
(a) Tool 1 (Drill).
(b) Tool 2 (Reamer).

reaming, tapping and milling without the need to pre-set the tools to a specified length.

The G code controls the Z axis of the machine by storing the length compensation value in a specific offset register for a particular tool. Offsets used in tool length compensation essentially represent addition or subtraction of small tool length differences by the controller. G45 code, for example, is the addition of a tool offset value.

(ii) *Tool Offsets for a Turning Machine*

Figure 8.13a shows, diagrammatically, a tool turret with a cutting tool in position (1). The position centre of the turret can be established in the same way as the milling machine quill. The position of the cutting edge of the tool then has to be determined relative to the turret origin. The distances a and b define the tool offset. When the turret is indexed anti-clockwise, through 60°, as shown in Fig. 8.13b, the parting-off tool (2), will be in position ready for cutting. The cutting edge of this tool will also have to be defined relative to the turret centre and the values a' and b' may well differ from a and b for tool (1).

To establish tool offsets for a simple turning machine, consider the following procedure:

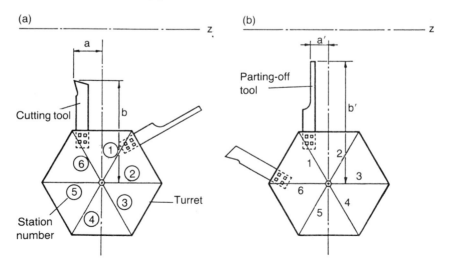

Fig. 8.13. Tool offsets for a turning machine.
　　　　(a) Tool 1 (Right hand cutting tool).
　　　　(b) Tool 2 (Parting off tool).

1. Select tool offset menu on control panel.
2. Enter reference tool number.
3. Touch tool on to end of workpiece (the Z zero plane in this case), using manual axis jog facility, and register this position. See Fig. 8.14a.
4. Touch tool on to workpiece or turn a diameter and register this position (see Fig. 8.14b).
5. Measure the diameter and key in the value in the appropriate units.
6. Repeat operations 1 to 5 for all the tools to be used.

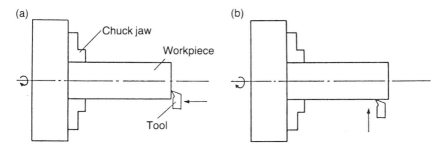

Fig. 8.14. Establishing tool offsets values.
    (a) 'Z' plane.
    (b) 'X' plane.

Under normal circumstances, tool offsets will be calculated for a range of tools and the machine, once set up, can be left to produce batches of components. However, in reality, tool wear takes place and tool breakages occur and these eventualities have to be catered for.

Compensation

Tool offsets alone are not sufficient to ensure satisfactory machining of components to fine tolerance limits. To bring about effective programming of a machine tool, various *compensating* factors have to be taken into account.

The main factors are cutter diameter compensation, tool nose radius (TNR) compensation and tool length compensation.

Compensation is achieved by the programmer or operator incorporating a compensation block in the part-program. The compensation block would perform the following functions:

1. To assign a specific register to a specific tool. The operator inputs the required compensation into the assigned register.
2. To simplify programming in terms of roughing or finishing operations. This is achieved by the operator inputting a radius value which is larger than the actual radius of the tool. The difference between these two values is the amount of stock left on the part after the current pass of the tool.
3. To allow a cutter of a different diameter, to that previously specified.
4. To define the direction of compensation, i.e. to the right or left of the part.

Cutter Diameter Compensation

This feature allows the positioning of the tool, prior to machining, to the left (or right) of the workpiece, by a distance equivalent to a specified tool radius. The part-program is written as if the tool radius were zero. The dimensions to be used for the cutter path will therefore become the actual dimensions of the component drawing, thus obviating the need for calculations.

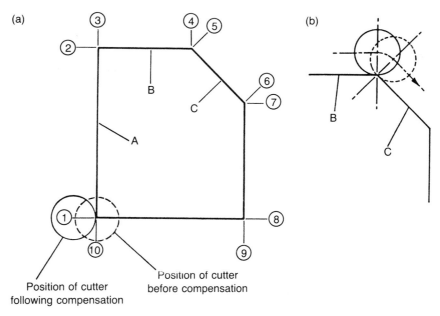

Fig. 8.15. (a) Diagram to show left cutter compensation.
(b) Centre point of cutter rotated about .4 and 5.

To explain cutter diameter compensation in more detail, consider the diagram in Fig. 8.15a. The component profile is to be machined in the direction indicated by the encircled numbers using an end milling cutter. A left cutter compensation code G41 will be used here.

The program will instruct the tool to move to point (1) but as the compensation factor is used the tool will move away from the workpiece in a direction 90° to edge A by an amount equal to half the tool diameter.

Consider the change in angle between edges B and C.

As cutting progresses it is essential that the direction of the cut is parallel with the desired finished profile — therefore the milling cutter will have to change its direction of travel and rotate its centre point around point (4) and (5) before cutting along edge C. See Fig. 8.15b.

This occurs in conjunction with a change in the direction of linear interpolation and is used to bring the tool radius perpendicular to the new surface to be machined. A G39 code is used to achieve this function.

Tool Nose Radius Compensation

The two cutting edges of a conventional turning tool appear to come to a sharp point. However, in reality, the point of the tool will be a radius, called the *tool nose radius*. Figure 8.16a shows a typical turning tool; the actual nose profile is shown in Fig. 8.16b. If the component shown in Fig. 8.17 were to be turned using this tool and the intersection of the

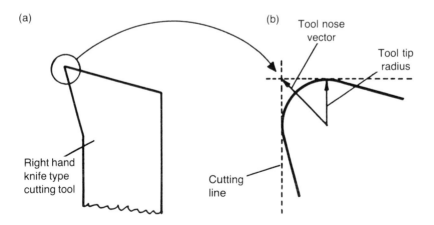

(a)

(b) Tool nose vector

Tool tip radius

Right hand knife type cutting tool

Cutting line

Fig. 8.16. (a) Typical turning tool.
(b) Close up of cutting edge showing tool nose vector and tool tip radius.

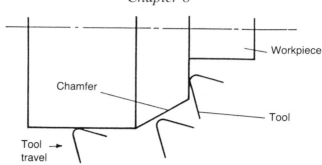

Fig. 8.17. Effect of tool path without TNR compensation.

cutting lines used as the distance of the tool from the workpiece, then as the tool travels in the area of the chamfer, no metal would be removed as this intersection is a small distance away from the actual tool tip radius. To overcome this problem a TNR compensation factor can be programmed which effectively moves the tool closer to the work by making the magnitude of the tool nose vector equal to the tool tip radius. As with cutter diameter compensation, the TNR compensation can be applied to the left or right of the workpiece. G codes G41 and G42 respectively.

The tool nose radius compensation may be cancelled after an operation by using the G40 preparatory function code. The use of this code has the effect of lengthening the tool nose vector to the intersection of the two cutting lines and hence moving the tool away from the work.

Tool Length Compensation

Tool length compensation is available on the more modern control systems and is used to compensate for different tool lengths. The major advantage of this facility is that it enables the programmer or operator to write a complete part-program without the need to know the exact lengths of the tools to be used. Tool length compensation can be away from the part (G43) or towards the part (G44).

Tool length compensation can be cancelled automatically for each tool before the next one is selected. The inclusion of a G49 code in the program achieves this cancellation.

Program Enhancements

As with conventional computer programs, part-programs for CNC machines can be enhanced by the addition of pre-programmed functions or by the use of subroutines. A few of the methods used for program enhancement are given below:

(i) *Canned Cycle* — A canned (or packaged) cycle is a set of pre-programmed instructions stored in the computer memory. The cycles are stored under a specified preparatory code address.

Typical canned cycles include tapping, drilling, boring, straight turning, taper turning, and facing.

(ii) *Automatic repeat cycles* — These are used for generating complex stock removal profiles. These repeat cycles are pre-programmed in the same way as the canned cycles mentioned previously and are known collectively as *fixed canned cycles*.

Stock removal cycles are very useful when several passes are required to remove stock. The result is a part outline made up of straight lines, slopes and arcs. For example G71 is a code which represents a longitudinal roughing cycle used in turning.

Figure 8.18 shows, diagrammatically, the roughing cycle for a

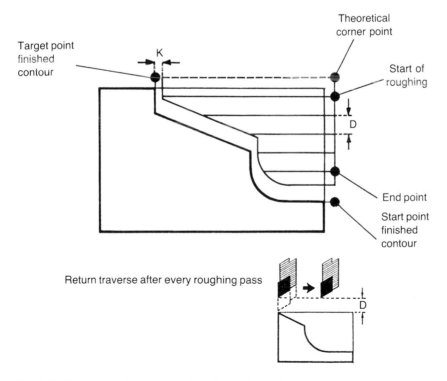

Fig. 8.18. Diagrammatic representation of roughing cycle for a turned component (reproduced by kind permission of Traub Limited).

turned component. The data block used would cause a number of roughing passes to be carried out to generate the desired profile, yet leave sufficient stock for finish turning.

The data block would have the form:

NO G71 A P Q I K D F S

where:

A, P and *Q* define the finished contour. *A* is the subroutine number if the contour is defined by a subroutine.

P and *Q* are block numbers if the contour is defined in the main program. (*P* = starting point, *Q* = target point).

I is the machining allowance towards finished contour in *X* direction.

K is the machining allowance towards finished contour in *Z* direction.

D is the depth of cut.

F and *S* are feed rate and spindle speed respectively.

Canned cycles enable part-programs for the generation of complex parts to be written efficiently.

Subroutine Programming

If instructions are needed more than once in a program, it is often convenient to write them as a *subroutine*. Subroutines are a very powerful time-saving device and provide the capability of programming certain fixed sequences and storing them in memory. They also provide the opportunity of creating specific canned cycles for a particular product or application. The subroutine may be accessed from any number of points in the part-program (or main program). After execution of the subroutine, control returns to the instruction following the one from which the subroutines was called. Subroutines are independent programs containing all the features of a stand-alone program and are treated by the control in the same manner as a part-program. Subroutines can be used for turning, milling and other CNC applications and are often referred to as *variable canned cycles*.

For example, consider block 30 in the section of main program, 004, shown below:

```
004    N10  _____
       N20  _____
       N30   P200   M98
       N40
       N120  M30
```

This instruction transfers control to the head of program 200, M98 being the miscellaneous code for a subroutine call.

The section of subroutine program shown below has an M99 instruction in its final block which returns control to block 40 in program 004.

```
200      N10  --------
         N20  --------
         N30  --------
         N40  --------
         N50         M99
```

Parametric Programming

This is an advanced CNC programming feature which allows machining contours to be described mathematically. A full description of parametric programming techniques is beyond the terms of reference for this chapter but suffice to say that a standard machining profile can be defined in terms of a mathematical expression (using combinations of variables, constants and operators just like a conventional computer program), and then assign values to variables to describe the profile dimensions.

Do Loops

A *Do Loop* (this term probably derived from FORTRAN) is a CNC programming feature allowing the efficient programming of repetitious machining operations which would otherwise require a large number of blocks. A do-loop can be thought of as an embedded subroutine. The codes for the execution of do-loops will vary from machine to machine; the section of part-program for a turning operation given below shows a typical use of a do-loop.

```
N08   G0    X17.5   Z1
N09   G0    Z-55    F100
N10   G01   X18     F200
N11   G0    Z1
N12   G81   FROM 8 TO 11    REP 3    X -1.5
N13   G0    X12     Z1
```

Block 8 instructs the tool to rapid traverse to position $X = 17.5$ mm and $Z = 1$ mm. Blocks 9 to 11 perform a single roughing cut. The G81 code in block 12 invokes a repeat cycle of blocks 8 to 11, inclusive on three occasions, where each time through the sequence the value of *X* is reduced by 1.5 mm. As the programmer may define any required shape, within machinability limits, do-loop programming is considered far superior to the canned cycle approach.

Entering the Program

Having written the part-program, the instructions need to be entered into the CNC system. There are a variety of ways of entering the program and the more sophisticated methods are described later in this chapter. The traditional method was to encode the program instruction on paper tape and then feed the paper tape into a reader on the CNC machine. Data can also be entered manually by putting the control into an MDI (Manual Data Input) mode. The program can then be entered one-line at a time.

8.5 PROGRAMMING CASE STUDY — TURNING

To explain how a part-program is constructed, a case study is given which shows how a program is built up in discrete sections. Line diagrams accompany the code to provide a visual interpretation of the program.

The Component

The component to be produced is shown in Fig. 8.19. This is a simple turned part, but it has features such as longitudinal surfaces, tapers, internal radii, and external radii in order to help demonstrate the concepts involved in programming turning machines.

The Machine

The machine used for this case study is the Easiturn 3, manufactured by Denford Machine Tools Limited, shown in Fig. 8.5. Easiturn 3 is a low-cost CNC lathe designed as an efficient production machine tool and also for training and re-training operators.

Setting Up

The component is to be turned from 38 mm dia. bar stock and is positioned in the machine as in Fig. 8.20. The origin of the machine axes

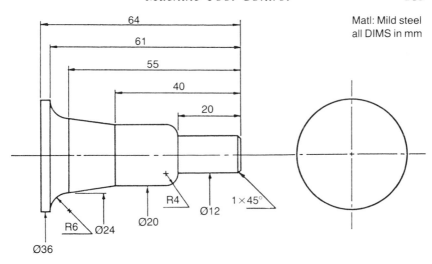

Fig. 8.19. Turned component to be produced for case study.

are also shown. Two tools are to be used for machining — tool 1 is a right hand knife tool and tool 2 is a 3 mm wide parting-off tool. The tools are positioned in the tool post as shown in Fig. 8.21.

Plan view

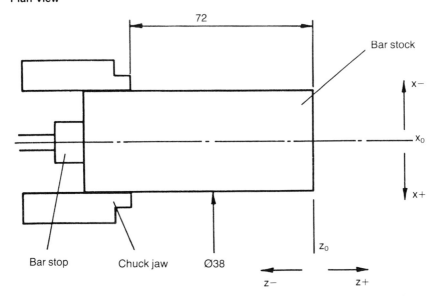

Fig. 8.20. Bar stock positioned in chuck.

Plan view

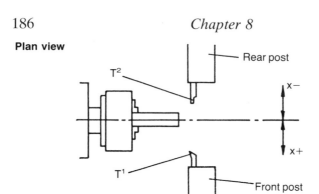

Fig. 8.21. Tool layout for case study.

The tools are firmly clamped and the tool offsets are set by positioning each tool in turn on the Z axis and then on the diameter, and keying in the values on the control panel.

The Program

Sections of the program will be listed and a description will follow. Diagrams will be used where clarity of expression is desirable.

N01	G00	X50	Z50
N02	M06	1	
N03	M03	1400	
N04	M08		

G00 instructs rapid traverse of the tool, clear of the workpiece, to a position 50 mm in the X direction and 50 mm in the Z direction, from their respective origins (see Fig. 8.22). M06 is a tool change command and tool 1 (in front tool post) is selected. M03 starts the spindle rotating in a clockwise direction at a speed of 1400 rev/min.

M08 starts up the coolant pump.

N05	G00	X21	Z0	
N06	G01	X −2	F120	
N07	G01	Z1	F348	

Block 5 is a rapid traverse of tool to a position 2 mm clear of the bar. Block 6 instructs the tool to maintain present Z position, travel in the X direction 2 mm beyond the bar centre line to face up the end of the work. The rate of feed to be 120 mm/min. Block 7 lifts the tool clear of the work.

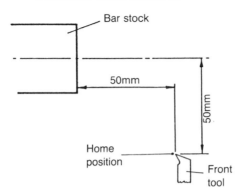

Fig. 8.22. Diagram shows Tool 1 in tool change position.

N08	G00	X17.5	Z1
N09	G01	Z −55	F102
N10	G01	X18	F198
N11	G00	Z1	

Blocks 8 and 9 produce a rough cut of 1.5 mm of metal to a length of 55 mm from the end of the bar. Blocks 10 and 11 lift off the tool and return it to position $Z = 1$ mm, to complete the cycle.

| N12 | G81 | FROM 8 TO 11 | REP 3 | X −1.5 |

Block 12 is a do-loop instruction (G81) which performs the roughing cycle a further 3 times to produce a diameter of 26 mm, as shown in Fig. 8.23a.

N13	G00	X12	Z1
N14	G01	Z −40	F150
N15	G01	X12.5	F252
N16	G00	Z1	
N17	G81	FROM 13 TO 16 REP 1	X −1

Blocks 13 to 17 perform a similar cycle to the previous one to produce a diameter of 22 mm along a 40 mm length. See Fig. 8.23b.

N18	G00	X9	Z1
N19	G01	Z −20	F102
N20	G01	X9.5	F252
N21	G00	Z1	
N22	G81	FROM 18 TO 21 REP 1	X −2

Blocks ⑧ to⑫

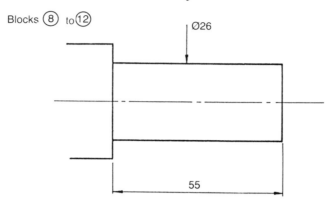

Fig. 8.23. Stages in Case Study: (a) Roughing cycle to produce 26 mm dia.

Blocks ⑬ to ⑰

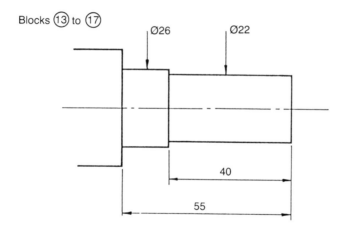

Fig. 8.23. Stages in Case Study: (b) Roughing cycle to produce 22 mm dia.

Again, blocks 18 to 22 form a small roughing cycle to produce a 14 mm diameter, 20 mm long, as in Fig. 8.23c.

```
N23    G00    X12
N24    G00    Z -39.5
N25    G01    X11    F150
N26    G01    X13    Z -55    F102
```

Blocks 23 to 26 initiate a roughing cut to produce the taper shown in Fig. 8.23d.

```
N27    G01    X18.5    Z -61
```

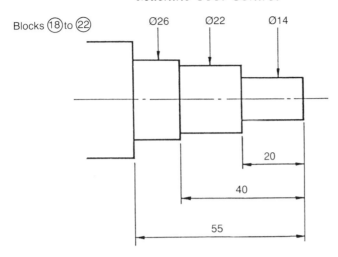

Fig. 8.23. Stages in Case Study: (c) Roughing cycle to produce 14 mm dia.

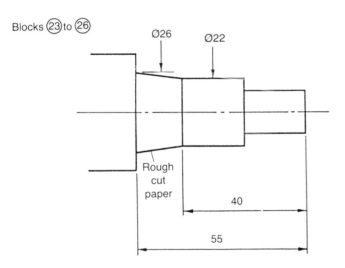

Fig. 8.23. Stages in Case Study: (d) Roughing cut to produce a taper.

Block 27 removes a stock of metal in preparation for the radius (Fig. 8.23e).

```
N28    G00    Z −19
N29    G00    X8
N30    G01    X7     F150
N31    G02    X11    Z −23    F102    XC7    ZC −23
```

Block ㉗

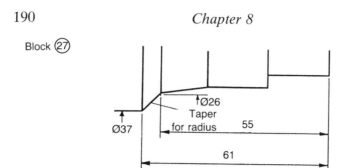

Fig. 8.23. Stages in Case Study: (e) Taper in preparation for radius.

Blocks 28 and 29 traverse the tool back in readiness for a rough cut of small radius. Block 30 positions the tool for the start of the cut. Block 31 has a G02 code which means circular interpolation in a clockwise direction. X11 and Z –23 give the final tool position, while X_c and Z_c give the co-ordinates of the centre of the arc, as shown in Fig. 8.23f.

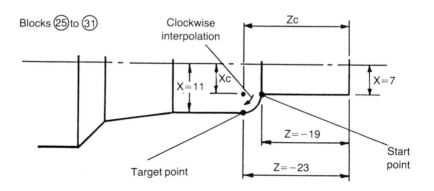

Fig. 8.23. Stages in Case Study: (f) Generation of arc.

N32	G00	Z1		
N33	G00	X5		
N34	G01	Z0	F	150
N35	G01	X6		Z –1

Blocks 32 to 35 instruct the cutting of the chamfer on the end of the bar. The tool is now in position to perform the final finishing cut (see Fig. 8.23g).

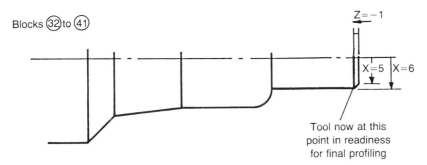

Fig. 8.23. Stages in Case Study: (g) Generation of chamfer.

N36	G01	Z −20			
N37	G02	X10	Z −24	XC6	ZC −24
N38	G01	Z −40			
N39	G01	X12	Z −55		
N40	G03	X18	Z −61´	XC18	ZC −55
N41	G01	Z −67			

Blocks 36 to 41 form a sequence of operations to turn the finished profile. The sequence can best be understood by reading each block in turn and referring to the diagram in Fig. 8.23h.

N42	G00	X50	Z50
N43	M06	2	
N44	M03	1200	

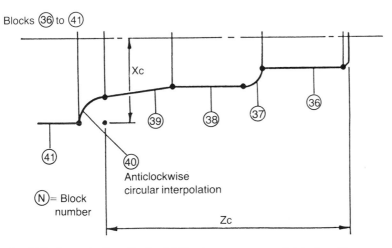

Fig. 8.23. Stages in Case Study: (h) Final cut to generate finished profile.

Block N42 returns the tool to the tool change position. M06 selects tool 2 (the 3 mm wide parting-off tool). M03 selects a spindle speed of 1200 rev/min. for parting off.

N45	G00	Z −67	
N46	G00	X −22	
N47	G01	X2	F60
N48	G00	X −30	
N49	G01	Z50	

Blocks 45 and 46 rapid traverse the tool ready for parting-off. Note that the X co-ordinate is negative as the tool is in the rear tool post. Block 47 feeds the parting-off tool until its tip is 2 mm beyond the centre line, to ensure satisfactory parting. The tool is then brought clear of the work by blocks 48 and 49. The finished component is shown in Fig. 8.24.

N50	M05
N51	M09
M02	

With the operation complete, the machine can be run down. These two final blocks close down the machine. M05 stops the spindle and M09 stops the coolant pump. M02 is the end of program marker for the control.

8.6 COMPUTER-ASSISTED PART PROGRAMMING (CAPP)

In the early days of NC and CNC, the parts to be machined were of two-dimensional configurations requiring simple calculations to describe the tool paths.

With the increased use of NC and CNC systems and the growth in the complexity of parts to be machined, the part-program is becoming more difficult to write and it is less easy for the part programmer to calculate the tool paths efficiently.

The specific limitations with manual part-programming include:

(i) Heavy use of manpower resources.

(ii) Loss of machine productivity.

(iii) In complicated point-to-point applications, manual part-programming becomes tedious and is subject to errors.

(iv) The codes used are not universal and can lead to confusion if a variety of machines are to be programmed.

Fig. 8.24. Photograph of machined component.

These limitations can be overcome by using the computer to assist with the part-programming process.

Computers can perform the required mathematical calculations quickly and accurately, and the computation errors which arise with manual calculations are reduced with *computer-aided part-programming*.

There are a wide variety of computer-assisted part-programming languages available that have been developed to perform most of the calculations that the programmer would otherwise have to do. The use of these languages saves time and leads to a more accurate and efficient part-program.

The machining instructions are written in English-like statements (rather like a high-level computer programming language), which are then processed by the computer to produce the machining codes.

The part-programmer defines the geometry of the workpiece and then specifies the sequence of operations and the path of the tool. The computer then translates the high-level instructions, calculates offsets and compensations, then post-processes the information for use by the machine tool.

Part-programming languages are designed to make it convenient for a part-programmer to input the necessary information so that the desired part-program can be prepared. The more common languages that have evolved include *APT, ADAPT, UNIAPT, SPLIT* and *CINTURN II.*

APT (Automatically Programmed Tools) is probably the most well-known language and includes statements of the form:

```
PT4 = POINT/2,3
GOTO/HOLE 6
GORGT/L2, PAST, L3
```

where:

PT4 = POINT/2,3 means that PT4 is a symbolic designation of a point whose X co-ordinate is 2 and Y co-ordinate is 3,

GOTO/HOLE 6 — This instructs the tool to move to the X and Y co-ordinates of a point called HOLE 6, which has been defined elsewhere in the program, and

GORGT/L2, PAST, L3 means start moving the tool to the right and then move the tool along a line called L2 until it passes a line called L3.

These high-level statements make it easier for the part-programmer to understand than conventional part-programs comprising combinations of G codes and M codes, thereby increasing machine tool programming efficiency.

8.7 LINKING CAD WITH CAM

Many modern manufacturing systems allow data created during the design phase to be exploited in the manufacturing phase. There has been a progressive merging of computer-based design and manufacturing facilities and it is a central feature of new CAD/CAM systems that they not only aid component design but also generate the necessary numerical control information to instruct the appropriate machine tools. In this way and other ways CAD/CAM systems are evolving to intervene *directly* in the manufacturing process.

Transforming Design Data into Machining Instructions

Having created a drawing data file within the computer that completely describes the component, this information can now be translated into a set of instructions to which the machine tool can respond. At this stage it is useful to make use of the drawing layer technique (explained in Chapter 4) to access separate geometric profiles for the production of manufacturing instructions. Having produced the component drawing, a processor creates a complete set of machining programs to produce the drawn component. This direct technique eliminates errors that are frequently introduced when drawing information is manually transferred onto cutter data sheets. Similarly, data can be transferred from a 3-D modeller to the manufacturing system.

The ability to link the CAD computer directly with manufacture is generally termed *Direct Numerical Control* (or DNC). DNC becomes particularly significant when two-way data transfer is available. Instructions generated in the CAD system can be directly down-loaded to the plant and machinery, minimizing or even eliminating the need for manual input or the use of unreliable shop-floor consumables, such as paper tape or magnetic cassette. Along with machine instructions, the details required to set up a job can be passed to the operator, often prior to the work being required so that preparations for the next job can start. The necessary material and tools can be selected, pre-setting onto pallets or loading carousels can be carried out and stortages identified. Early identification of problems will result in increased utilization of equipment.

8.8   CAD/CAM LINK — A SIMPLIFIED STUDY

To demonstrate the way in which drawing data can be transposed into machining instructions, a simple case study is given.

The study describes the use of various software elements of the Micro Aided Engineering system outlined in Chapter 4.

The object is to translate the drawing data for the flat plate (Fig. 8.25) into instructions for use by a milling machine.

Procedure

1. Load MAEDOS drawing routine and re-call the drawing which was saved as a symbol (Chapter 4). By selecting the relevant parts of the drawing, using the joystick and hit keys, the profile and point data is stored for subsequent machining. (The system allows cutter paths to be automatically locked onto these profiles with desired offsets.)

   There are a number of user-definable line types and these are used to establish the component geometry for machining by giving each contour or pattern of holes unique line types. Line type 50 is

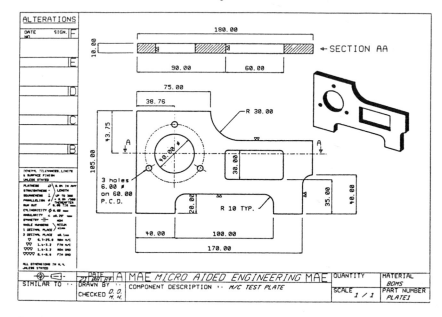

Fig. 8.25. Drawing of component produced in Chapter 4.

selected and the component origin is defined.

Line type 51 is used to define the centre of the three small holes.

Line type 52 defines the centre and radius of large hole.

Line type 53 defines the slot profile.

Line type 54 defines the outer profile of the plate in an anti-clockwise direction.

The geometry definition is now complete and the data is saved on disk.

2. Load MAELINK software. This links the MAEDOS draughting system to the MAECAM graphical NC programming system. It enables the NC part-programmer to take geometry data direct from a MAEDOS drawing and use it on the MAECAM manufacturing system.

A question and answer routine regarding machine set-up details now takes place, in which the user enters the following information:

Machine type (for the post-processor),
Machine origins,
Workpiece loading position, and
Tool change position.

A tool file is then displayed for the user to pick off the required tools for the job. For this example the tool details are:

Tool 1 = 6 mm drill
Tool 2 = 3 mm end mill
Tool 3 = 5 mm end mill

3. Set up details for the tool path simulation program. Geometric data is retrieved from the MAEDOS file and additional machining statements are given, e.g. drill depth, spindle speeds for each tool, feed rates and so on.

An interactive question and answer session enables the tool actions to be specified by matching each tool with the line type specified earlier to ensure that the drill drills the hole and the end mills cut the profiles, etc.

The machining statements are complete and the part-program is automatically generated. There is an automatic exit from MAELINK at this stage.

4. Load MAECAM software. MAECAM is a graphical NC programming system using CONPIC software. It gives graphical verification of the geometry, tool, clamp location, and tool path to reduce proveout time. MAECAM also carries out syntax checks and effects the post-processing of part-program into machine language.

Having completed the syntax check, the tool path simulation can take place. By selecting each tool in turn, the path of the tool relative to the workpiece is replicated on the screen. Figure 8.26 shows the path of the tools when cutting each separate contour or pattern.

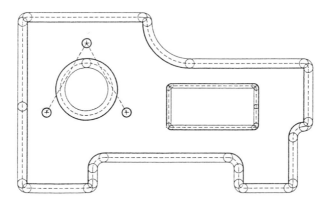

Fig. 8.26. This figure shows the path of the tools when cutting each pattern or contour.

When satisfied with the tool path, the post processing can begin. The MAEGEN post process generator is used to convert the part-program into a series of G codes and M codes etc., which can be viewed on the screen or, alternatively, be written out to cassette tape or to a printer.

5. Load MAESIM software. MAESIM provides graphical simulation of the overall tool path on *X, Y* or *Z* planes and isometric views. It can also edit part-programs with instant verification and gives fast tape prove out for experimental programmes without tying-up shop floor facilities. Figure 8.27 shows the simulated overall tool path for the milled component.

The CAD/CAM link is of increasing importance to today's manufacturing industry. It is estimated that 90% of industry in the UK is based upon batch manufacture. Therefore, intelligent use of the CAD/CAM link between drawing office and shop floor can produce enormous benefits in terms of response time and reduction of work in progress.

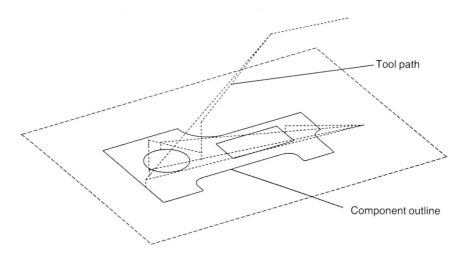

Fig. 8.27. Simulated overall tool path for the milled component.

Chapter 9

Robotics in Computer-Aided Manufacture

9.1 INTRODUCTION

The application of robotics to Computer-Aided Manufacture has been increasing steadily over the years as management realizes the potential of industrial robots. Robots can be easily applied to many areas of manufacture, including assembly, welding and spraying, and as a result more and more companies are using robots to aid their manufacturing activities. A robot is also an essential element of a flexible manufacturing system.

This chapter explains what a robot is by describing the mechanical features of a robot and the senses that they have.

The methods of programming a robot are discussed and typical software systems are outlined. Many examples are then given of the different industrial uses of robots and finally artificial intelligence is introduced.

9.2 WHAT IS A ROBOT?

History and Development

The word ROBOT had not been invented before 1920 and for the following thirty years all robots were fictional. The word Robot entered the English vocabulary with the translation of the Czech writer, *Karel Capeks*, play, *Rossums Universal Robots*, in 1923. Robot is derived from the Czech word *ROBOTNIK*, which means worker.

The image of a robot for most people has largely been conditioned by science fiction films and a popular conception of a robot is a tall mechanical humanoid figure filled with electro-mechanical components, capable of performing human actions, albeit clumsily, whilst displaying sub-human intelligence.

A small number of robots were built in the 1940s and 1950s but they were largely research toys used for demonstration purposes, and it wasn't until the early 1960s that robots began to be employed to do useful work.

199

These robots could mimic the human arm by swinging up and down and from side to side, reaching out and gripping. The majority of these machines had mechanical joints which closely resembled the joints of the human arm but the rest of the machine bore no resemblance to a human being. The robot arm was in fact mounted on a supporting structure.

This type of robot is capable of repeatedly following a pre-set cycle of instructions without error. These machines have no intelligence and are termed *first generation* robots. There are currently thousands of this type of robot working in factories throughout the world.

Robots which have advanced computing power, giving them the ability to sense their environment, learn from a program or from their own experience and to make adjustments to their activities, are being developed in the advanced manufacturing nations.

This type of robot is referred to as a *second generation* robot, as it possesses a degree of intelligence. However, the mechanical appearance differs very little from the first generation machines. The second generation robots have a certain independence of action over and above what they have been programmed to do. This is brought about by the increased use of sensors which can identify actions and feedback information for analysis.

Definition of a Robot

Robot is a general term for a machine which is capable of performing human-like actions and functions, without the necessity to have human appearance. The description of a robot may also be applied to automated machines, particularly those capable of carrying out mechanical functions which can also be performed by human beings.

The writer *Isaac Asimov* in the 1940s took it upon himself to define robotics in simple terms. He did this by proposing three laws of robotics to which all robots should conform.

These were:

1. A robot must not harm a human being, nor through inaction allow one to come to harm.
2. A robot must always obey human beings, unless that is in conflict with law 1.
3. A robot must protect itself from harm, unless that is in conflict with the first two laws.

There are countless applications for which a robot may be used, but this chapter is only concerned with the type of robot used in manufacturing industries. A typical manufacturing robot consists of three elements: the mechanical structure, the control system and the power unit.

(i) *Mechanical Structure* — This encompasses the main frame, trunk or pedestal and the associated linkages which move an arm through a variety of axes providing a degree of versatility. The tools at the extremities of the linkage also form part of the mechanical structure.

(ii) *Control System* — There are many types of control systems such as pneumatic logic, electronic sequencers and microprocessors. In a general sense robots can be non-servo or servo-controlled. Non-servo robots have mechanical stops and limit switches for positoning, whereas servo-controlled robots have sophisticated feedback devices which monitor the appropriate variables until a point is reached at which the controller receives a signal instructing the termination of actuation of the robot arm. A servo-controlled robot can be of the point-to-point type (i.e. no control of path between points to be reached) or continuous path controlled (controllable smooth continuous movement between points).

(iii) *Power Unit* — All robots require power to operate their mechanical systems. Power units are usually hydraulic, pneumatic or electrical. The typical breakdown of the use of different power systems for manufacturing robots is shown below.

System	Percentage of Robots using System
Hydraulic	50
Pneumatic	30
Electrical	20

A large number of robots, however, use combinations of two or more of the sytems.

Figure 9.1 shows the three elements of a robot.

Types of Robot

The first generation manipulator type robots can be classified into four distinct groups.

Viz: Fixed sequence robots

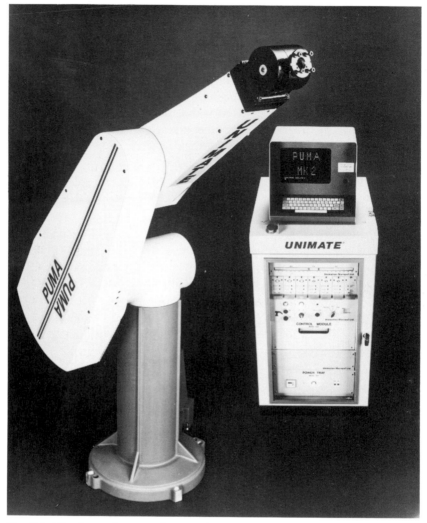

Fig. 9.1. The three elements of a manufacturing robot: mechanical structure, control system, and power unit (courtesy Unimation (Europe) Limited).

Variable sequence robots
Numerical control robots
Playback robots

(i) *Fixed sequence robots* — These are machines capable of performing successive steps of a given operation in a pre-determined sequence. The pre-defined informaton is usually built into the robot and cannot be readily changed. In other words a fixed sequence robot is designed for a specific operation.

(ii) *Variable sequence robots* — This type of robot is similar to the fixed sequence machine with the exception that the operating information can be changed. The simplest form of variable sequence robot is one in which the operating information is changed by altering the positions of plugs on a plug board. However, the more modern versions of variable sequence robots use microprocessors to define their operating regimes.

Fixed and variable sequence robots are often referred to as *pick-and-place* robots although they can be used for other types of task.

(iii) *Numerical control robots* — This type of robot has the same mode of control as an NC machine. The sequences, condition and positioning commands are effected by numerical data fed into the machine. The command programmes, as with conventional numerically controlled machines, can be in the form of punched tape, magnetic tape or similar input media.

(iv) *Playback robot* — One characteristic of fixed and variable sequence robots is that they are difficult to re-program. This is due to the fact that the control system and memory are all embodied in a complex set of limit switches, interlocks, stops and electrical connections. Not only does this type of electromechanical arrangement prove tedious to change but it also limits the number of different sequence steps that can be accommodated within the control system. A playback robot overcomes many of the programming problems associated with the fixed and variable sequence types.

A playback robot incorporates a memory and has the capacity to be taught a sequence of movements required by the human operator. When the robot has been taught a sequence of operations it holds these instructions in memory for recall when required. The instructions are played back through the control system and the robot meticulously replicates its taught sequence. The teaching process is carried out by the operator using the controls to drive the robot limbs to the required position for each operational step and then to record the exact position of the robot by pressing a button on the controller. The robot is moved to the next position in the operation and the position recorded.

This procedure continues until the operation is complete and all the movements are recorded in the memory.

This discrete-step form of teaching is called *point-to-point* control where the robot is instructed to move from point to point on a start and stop basis, and perform a given function at each stop point. This system is ideal for applications such as spot welding where a rapid movement followed by a pause is acceptable.

There are, however, applications in manufacturing industry where it is necessary to control not only the start and finish positions of each step, but also the path traced by the robot hand as it travels between two extremes. An example of this requirement would be a seam welding procedure where a robot has to hold a welding gun and traverse it along a complex contour at the correct speed to produce a neat, strong weld. This type of robot requires a memory that is sufficiently large to allow a *continuous path* to be recorded.

Robot Configurations

Robots in manufacturing industry come in a variety of shapes and sizes. They are capable of various arm manipulations and they possess different motion and control systems. However, almost all industrial robots have one of four basic physical arm configurations.

1. *Cartesian* — The main frame of a Cartesian co-ordinate robot consists of three orthogonal linear sliding axes, as shown in Fig. 9.2a. The operating mode of this type of robot can be thought of as being similar to that of a CNC machine tool.

2. *Cylindrical* — The main frame of a cylindrical co-ordinate robot consists of a horizontal arm mounted on a vertical column which, in turn, is mounted on a rotary base, as shown in Fig. 9.2b. The horizontal arm moves in and out, the carriage moves up and down on the column, and these two units rotate as one unit on the base.

3. *Spherical* (or Polar) — A spherical co-ordinate robot consists of a rotary base, an elevated pivot, and a telescopic arm which moves in and out, as shown in Fig. 9.2c.

4. *Angular* (or Revolute) — This type of robot consists of rigid members connected by two rotary joints and mounted on a rotary base, as shown in Fig. 9.2d. This configuration closely resembles the human arm. Its accuracy is not as good as the three previous systems as

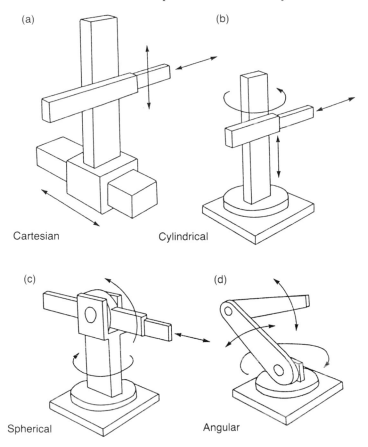

(a)

(b)

Cartesian

Cylindrical

(c)

(d)

Spherical

Angular

Fig. 9.2. Basic robot arm configurations.

errors are accumulated at the joints. However, it can move at high speeds and has excellent mechanical flexibility.

9.3 MECHANICAL ELEMENTS

There are a variety of robot types on the market with a wide range of mechanical elements. However, the majority of these designs have a pedestal or trunk to which all the other robot limbs are directly or indirectly connected. The pedestal provides a solid base from which the limbs can operate. The pedestal may be firmly secured to a rigid struc- ture, such as a floor or wall, or can alternatively be mounted on a moving platform. The pedestal or trunk usually has a waist joint allowing the body to rotate about the pedestal centre line (see Fig. 9.3). An arm is usually attached to the upper half of the trunnk and a hand attached to the arm at a wrist provides the required range of movement.

Arms

Arms are a key factor in the design of a robot as they position the working hand in the proximity of the work. The simplest type of arm is

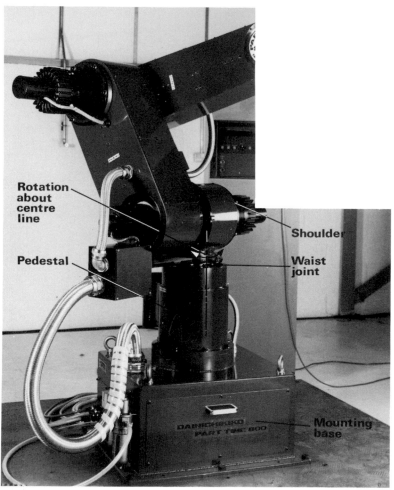

Fig. 9.3. Robot assembly showing trunk rotation about pedestal axis, via waist joint (courtesy Dainichi-Sykes Robotics Limited).

one which slides backwards and forwards relative to the trunk, as shown in Fig. 9.4a. This type of arm, as well as the in/out sliding motion, can also be raised, lowered and rotated about the vertical trunk. Figure 9.4b shows the typical coverage range for this type of arm. The more sophisticated type of arm is one which has two sections connected by joints, as shown in Fig. 9.4c. The mechanical design of this type of arm is similar in nature to the human arm in that it possesses shoulder and elbow joints. However, the robot arm is more versatile than the human arm as the mechanical joints are double jointed and therefore give the robot arm a far greater range of movement than a human arm. The total area covered by the robot arm and body movement is called the *work envelope*.

Arm movement

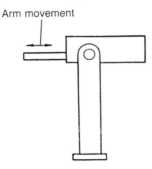

Fig. 9.4 (a) Simple arm configuration.

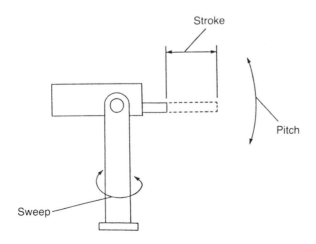

Fig. 9.4 (b) View showing pitch, sweep and arm stroke.

Chapter 9

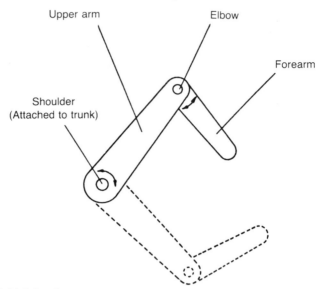

Upper arm Elbow

Forearm

Shoulder
(Attached to trunk)

Fig. 9.4 (c) Jointed arm configuration.

Hands

The robot hand is the device which facilitates the grasping action of the robot. The hand is connected to the robot arm by a wrist joint, providing rotation of the hand at 90° to the arm. Hands (or *end effectors*) can be divided into two categories: grippers or tools.

Grippers

Grippers act like mechanical fingers and are used for gripping, holding and releasing objects. The choice of a suitable gripper for a particular application is vital as it is this point where the robot interfaces with the work. The cost of a gripper can be as much as 20% of the overall cost of the robot. The way in which a job is handled or grasped largely depends upon the nature of the material to be handled.

Many forms of gripper exist, the more common ones being: mechanical, vacuum cup and electromagnetic types. Typical mechanical gripper types include:

Self aligning finger pads — used to ensure a secure grip on flat sided components.

Multi-size pickup — This type has a number of different sized insets, for use where a variety of predetermined component shapes and sizes are to be handled. The robot can be programmed to ensure that the correct inset matches the component to be handled.

Wide opener — Used when the components to be handled are not always in the same orientation. As the gripper closes it sweeps the component into its grasp.

Movable jaw — Used when access can be gained to the underside of a component. Simplicity of design makes this one of the most economical types of gripper.

Figure 9.5 shows diagrammatically the various types of mechanical gripper.

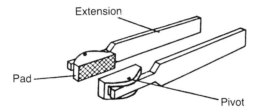

Fig. 9.5 (a) Self aligning end-effector.

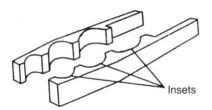

Fig. 9.5 (b) Multi size pickup end-effector.

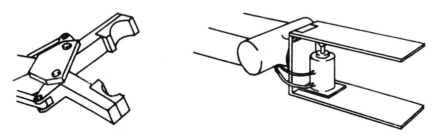

Fig. 9.5 (c) Wide opening jaw end-effector.   **Fig. 9.5** (d) Movable jaw end-effector.

Vacuum cup gripper systems normally employ cups made of an elastic material which conforms to, and forms a seal with, the surface of the part to be handled. To achieve the best utilization of a cup the largest possible vacuum (or pressure differential) should be used, although to obtain a fast handling system a large cup with a lower vacuum is used. The vacuum will not form until the cup has sealed on the component; therefore, to increase the speed of a vacuum system it is advantageous to mount the cup on spring-loaded rods and program the robot so that the cup makes contact with the component long before the arm reaches its final pick-up position.

The main attraction of vacuum grippers is that only one surface of the component has to be used. They also provide even surface pressure distribution and are cheap and reliable. Vacuum grippers are particularly suitable for handling sheets of glass and similar materials. Figure 9.6 shows a vacuum gripper arrangement.

Magnetic grippers could be considered if the component to be hand-

Fig. 9.6. Vacuum gripper arrangement.

led has a high ferrous content. Magnets can be designed and manufactured in numerous shapes and sizes to perform various tasks.

Electromagnets or permanent magnets may be used for grippers. Electromagnets are preferred since the component can be released from the gripper by switching off the electric current, whereas a component would have to be mechanically ejected from a permanent magnet gripper.

Tools

For certain assembly and manufacturing tasks a robot can be equipped, with a tool in place of a gripper. The tool can be mounted directly onto the wrist or can be located in a tool holder attached to the wrist.

There are many types of tools suitable for use by a robot, the most common ones being:

Spot welding gun.
Arc welding torch.
Socket (for running down and tightening nuts).
Paint spray gun.

Robot Movements

For a robot to perform even the simplest pick-and-place transfer task the hand must be capable of moving in at least two different directions, i.e. it must have two *degrees of freedom.*

Combinations of movement of the robot joints (waist, shoulder, elbow and wrist) give the robot more degrees-of-freedom providing a versatile range of movement. The minimum requirement for an industrial robot, if it is to have flexibility, is six degrees-of-freedom. These are:

1. Waist rotation (arm sweep).
2. Shoulder swivel.
3. Elbow rotation/extension.
4. Wrist pitch.
5. Wrist yaw.
6. Wrist roll.

Figure 9.7 illustrates the six degrees-of-freedom of an industrial robot.

To make all operations possible a robot would require eight, nine or even ten degrees-of-freedom. This is to include robot body translation and gripper operation.

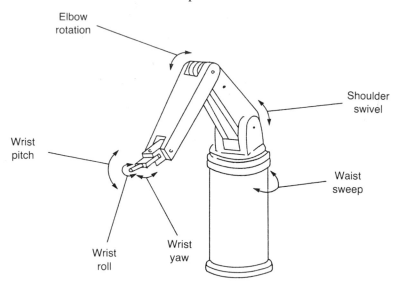

Fig. 9.7. Six degrees-of-freedom of an industrial robot.

Actuation Methods

Having described the basic mechanical elements of a robot the methods of actuation of these elements can now be described.

Whatever form of actuation is adopted there are certain requirements which must be satisfied.

The actuator should have a high power output to weight ratio and should provide the most flexible form of control of movement possible. Output torque should vary smoothly over the whole operating range, to guard against spasmodic movement. Finally, acceleration and deceleration, for the majority of tasks, should be as rapid as possible.

A method of actuation is required for each robot articulation. In addition to moving the body, arm, wrist and hand, most types of gripper also require a mechanism to provide hold and release functions. Actuation may be provided by electrical, pneumatic or hydraulic means and many systems use combinations of these.

(i)   *Electrical* — Electric motors can be a simple means of actuation, when coupled with gears or ball screws to provide a complete actuation system. The electric supply can be obtained from the mains or, alternatively, from battery packs. The majority of electrical robot actuation systems are d.c.

The disadvantage of using electric motors is that the power availability is somewhat limited unless large motors are employed. Another factor is that electric motors are relatively expensive and they also become hot with continuous operation.

(ii)   *Pneumatics* — these systems are confined mainly to fixed and variable sequence robots. Pneumatic actuation systems, due to their use of compressed air, have the merit of being cheaper than other methods, and their inherent reliability means that maintenance costs can be kept low. Many factories have a compressed air ring main providing a power source at a number of locations. Movement of robot limbs is provided by linear or rotary actuators and air motors.

Pneumatics are particularly suitable for the smaller robots, but unfortunately the system does not allow easy control of speed or position as air is a readily compressible fluid. However, low-power solenoid valves can be used to interface the pneumatics with micro-electronics allowing them to be driven from microprocessor control systems with mi 'mal power amplification.

(iii) *Hydraulics* — A hydraulic system is capable of providing high power outputs from small units and of producing linear and rotary movements. Hydraulic drives can be divided into actuators and motors. Actuators can be of the linear piston or rotary vane types, whereas the choice of hydraulic motor can be made from piston, gear, vane or ball types. The choice is largely determined by the application requirements. The piston actuator is the most preferred of all hydraulic drives since it is simple, cheap, very reliable and can provide push/pull movements without gearing.

Hydraulic systems generally have better response capabilities and are easier to control than electrical or pneumatic systems, which is the main reason why hydraulics are the most popular robot actuation system. There are, of course, certain drawbacks with hydraulics. One disadvantage is that the hydraulic fluid has to be pressurized, which means a pump has to be driven, usually by an electric motor. Another problem is that oil can escape through seals on poorly maintained systems, although as seal technology advances this is becoming less and less of a problem.

9.4 ROBOT SENSES AND SENSORY SYSTEMS

The simpler types of robot, such as the first generation pick-and-place machines, perform their tasks in a straightforward repetitive manner. These robots have no way of sensing the surrounding environment and are, to all intents and purposes, *unintelligent*.

If a robot is able to sense its environment (such as detection of absence or presence of a component, or orientation of a component) then the accumulated data can be used to control its behaviour. The key features of a second-generation robot are that it should possess the senses of sight or touch. The extent to which it needs to see or feel depends largely upon the tasks it is designed to perform.

Vision

Robots were first built with vision systems in the mid 1970s, when black and white television cameras were used to provide primitive vision capabilities. The simplest vision systems are based on a photoelectric cell, where the amount of current flowing through a circuit is proportional to the amount of light entering the photocell.

In this way it is possible to make a robot search for a light source by detecting the amount of light present, by the current flow, then rotating the body through a small amount and re-evaluating the current flow. If the current has increased the robot is rotating in the correct direction; if the current has decreased the robot changes direction. In this way the robot can be made to home-in on a light source.

Robot vision research has advanced considerably since the mid 1970s and now robots fitted with sophisticated vision systems are being used in manufacture and assembly. For example, Fairey Automation System have a vision guidance arc welding robot system. The system contains a welding torch with a leading and trailing camera, and four lasers which produce two structured light strips. Control is provided by two microprocessors and a video processor for real-time processing of the camera image. Sophisticated software provides full path control and handles all necessary image analysis in real-time. The vision system enables the welding gun to track complex three-dimensional seams giving seam errors of less than ± 0.5 mm.

Most of the vision systems used on assembly robots are based on *pattern recognition techniques*. The pattern of an object can be digitized and stored in the vision system memory. The robot camera can then assess a selection of objects, on a conveyor for example, and compare the pattern with its stored master. These systems enable object types and orientations to be recognized.

The more advanced robot vision systems include: infra-red cameras for seeing in the dark; thermal cameras for detecting the heat of objects; and material recognition cameras using a spectrometer.

Touch

The sense of touch can be built into a robot by providing it with tactile sensors. Tactile sensors fall into two broad categories — those which detect contact and those which not only detect contact but also sense the degree of contact (i.e. the amount of grip). Simple touch sensors usually employ a touch-sensitive switch and mechanical feelers. A microswitch stops movement of the hand when a mechanical stop is contacted.

The more complicated type of tactile sensor usually incorporates a force sensor to detect the amount of grip. The degree of grip is measured as a force which generates a control signal proportional to the force. These sensors can produce very accurate feedback enabling delicate operations, such as picking up an egg, to be performed. Force sensors are transducers which convert mechanical force into electrical signals. Strain gauges and piezo-electric devices are common force transducers. Some tactile sensors can detect over grip (i.e. gripper fingers closing in the absence of a component).

Tactile sensors with computer control have been developed for a wide range of industrial applications, in particular assembly operations, where accurate positioning of components is essential. To enhance the robot's potential, it is desirable to combine the different senses. One sense will be able to validate the findings of another, for example to test whether the vision system is seeing what the robot hand is feeling.

Hearing

One other important sense in robotics is that of hearing. There are circumstances where it would be useful for a robot to sense particular sounds in an industrial environment. Some sounds may signify danger and others may serve as commands. Voice recognition systems are becoming increasingly popular for robots as research into speech recognition advances. Robots which respond to simple spoken commands are now commonplace. These robots employ a microphone to detect the spoken word, which is then encoded, stored and decoded to perform a useful function. These systems use one-syllable words, i.e. one word produces a response in a given sequence and some action is taken, the next word produces the next action in the sequence and so on. The fundamental words used include: GO, RIGHT, LEFT and STOP. To make a robot more intelligent, additional senses may be employed.

9.5   PROGRAMMING AND CONTROLLING THE ROBOT

The operational versatility and accuracy of an industrial robot is determined by its programmability and controllability. For example, can the robot perform complex sequences of tasks, and can it readily change from one kind of operation to another? Other important considerations include:

   The robot's ability to make decisions,
   The behaviour and mobility of the gripper and arm, and the
   Time response of mechanical joints and limbs.

   The development and enhancement of second generation robots is dependent upon software and control capabilities.

Programming

Four basic programming techniques exist:

   (i) *Physical set-up* — This type of programming is used for the fixed and variable sequence robots, in which programs are set up by physically setting stops and switches, arranging wires, inserting punched cards or by using plug board methods. The physical set-up method has obvious limitations, yet it is ideal for programming small pick-and-place machines.

   (ii) *Walk through* — Used for robots which have magnetic tape, disk or minicomputer memory. The robot is moved by the operator through a continuous path, the motions being recorded in memory for future repetition. Figure 9.8 shows an operator programming (or teaching) a paint-spraying robot, using walk through techniques.

   (iii) *Lead through* — In which the programmer uses a control box, comprising joystick and keys, to command the various robot axes to move. Using this method the programmer can lead the robot through a cycle of operations, the various movements and conditions being recorded in memory for future use.

Fig. 9.8. Operator programming a paint spraying robot using the walk-through method (courtesy Binks-Bullows).

Figure 9.9 shows the programmer leading the robot through its operating cycle using a control box.

Speech recognition systems also allow programming facilities for some robot types.

(iv) *Downloading of operations* — This is where instructions are transmitted from a central computer, as in an FMS system. There is great benefit to be gained from this method of programming, as information from a solids modeller, for example, can be used directly by the robot.

Software

Research is being conducted into software systems in an attempt to make robots more reliable, versatile and intelligent. *High-level languages* are being developed to ease the programming of robots.

Programming a robot by walk-through or lead-through methods is a non-textual approach as the operator does not have to write a program. Operators without programming knowledge can easily program a robot

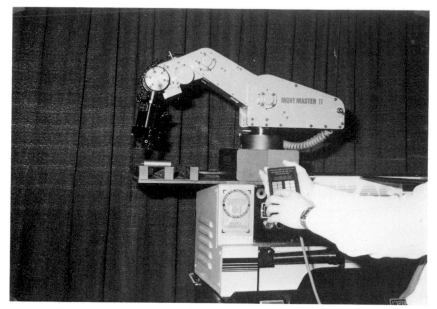

Fig. 9.9. Operator programming a robot using a control box.

in this way. However, the major disadvantage of these methods is that the program is inflexible and it is not possible for decisions to be made.

Textual programs have been developed which introduce the desired flexibility and which allow decisions to be made and branching operations to be carried out. These explicit programs require a precise instruction for every action the robot has to take (not unlike the programming of a CNC machine tool). Typical programming languages include *AUTOPASS, LANA* and *MAL*.

MAL allows a user to describe a sequence of steps necessary for an assembly task. Typically, robot programs can be written into memory in about four minutes. If it becomes necessary to change a command or correct a mistake, the memory location can be re-addressed and new data entered. PROM semiconductor memory allows ease of programming.

Control

Robot control relates to the positioning and response of mechanical joints as the robot performs its task. In particular, it is essential that the key robot elements, force and motion, are predictable and controllable.

Robot motions for assembly tasks, for example, require that the path traced by the hand is clearly defined and well understood, and that the hand does not foul anything. This can be checked using a solids modeller in more advanced systems. Movements of robot joints and limbs are mechanical systems in their own right and problems can arise due to inertia effects and backlash.

For example:

Suppose there is a requirement to rotate an arm through exactly 40°. A command signal is given to start the arm rotating and a signal is given to stop the arm when this position is reached. The arm will not be able to stop exactly at 40° as the inertia will carry the arm through and overrun its desired position. The arm will then have to be brought back to its exact position in steps. This is called an open-loop system (see Fig. 9.10).

For accurate positioning of a robot limb, a closed-loop system would have to be employed. A feedback signal is used to control the positioning of the limb.

As rapid acceleration and deceleration is required to make a robot device efficient, it is also important that the movements of limbs are dampened. If the system is underdamped the arm will oscillate about its target point until it finally comes to rest. If overdamped the arm will be slow in reaching its final destination.

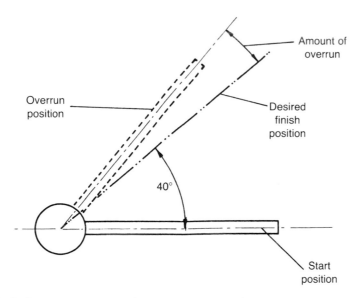

Fig. 9.10. Overrun of robot arm in an open-loop control system.

The damping factor is a critical element in robot control systems and the amount of damping should be such that the limb moves to its target position without appreciable oscillation in minimum time.

The requirement for more versatile robot operations has stimulated the research and development for more sophisticated control systems. Minicomputers and, more recently, microcomputers have provided the brain power for robots in industry. Computer-controlled robots are now working effectively in many manufacturing environments.

9.6 THE USES OF ROBOTS IN COMPUTER-AIDED MANUFACTURE

Robots in manufacturing industries are designed to be practical machines. Practical robots used to be confined to simple production jobs, such as welding and pick-and-place transfer operations. Now they perform many more operations, such as machine feeding, stamping, assembly, spraying, glass handling, plastics moulding, adhesive applicatios, gas detection, forging, casting, and warehouse automation. Robots designed to perform these specific tasks can be supplied by manufacturers. An assembly robot, for example, would have the necessary degrees-of-freedom of the wrist and the type of gripper to perform the majority of assembly tasks and it would be uneconomic to design an assembly robot from basics. There are, however, a few manufacturing tasks which still require one-off designs.

The following section briefly describes the main application areas for manufacturing robots.

Mechanical Handling and Palletizing

Every factory has a requirement for materials handling systems to some degree. It may just be a simple conveying task between machining stations or it could be more complex, requiring special process equipment such as swarf removal, CNC gantry loaders, or in-process buffer stores; it may be to lift a heavy or awkward component, or it may be to transfer delicate parts.

Mechanical handling robots can perform the traditional semi-skilled labour tasks in the factory. They can handle heavy objects and work in areas which may be unpleasant or unsafe for a worker. Work-handling robots are also particularly useful for loading and unloading conveyors in a flexible manufacturing system.

There are many types of work-handling robots but one of the more useful and versatile types is the *gantry robot*.

Gantry robots enable a wide variety of tasks to be performed, often over an area outside the work envelope of conventional robots, and where floor space is at a premium. Gantry robots can usually be floor-mounted, machine-mounted, suspended from structural supports or any combination of the three. Figure 9.11 shows a typical gantry robot.

Fig. 9.11. Gantry work handling robot (courtesy Fairey Automation Limited).

In manufacturing industries, the stacking of components on pallets is a means of automatically building up a load for removal to the next stage of manufacture, or to the despatch area or warehouse. It is much more efficient to move a single pallet loaded with components than it would be to pick up and move each component separately, particularly when a component needs to be properly orientated in order for the pick-up device to grasp it correctly. *Palletizing* robots are extremely useful where a large number of like components, preferably of regular shape, are to be loaded for movement about the factory floor. They are also particularly suitable for use in FMS.

Robots can be programmed so that the components are loaded in such a way as to ensure optimum pallet loading, i.e. the maximum number of components that can be safely loaded onto a pallet within the maximum specified load.

The choice of gripper is also important for a palletizing robot, as the gripper will have to suit the application. Figure 9.12a shows a robot loading a pallet with boxes. The gripper can be seen in Fig. 9.12b.

Fig. 9.12 (a) Robot loading a pallet.

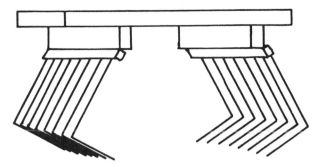

Fig. 9.12 (b) Special gripper for handling boxes.

Assembly

Assembly robots are, generally speaking, more sophisticated than pick-and-place robots. In particular, they may have five degrees-of-freedom with particular attention being paid to the dexterity of hand and wrist and positioning accuracy.

As second generation machines become more intelligent, it is possible to automate the assembly of complex components such as alternators and printed circuit boards. Assembly robots have been proved to be extremely efficient when set-up to work on assembly lines where the robot is statically mounted and the work passes on a conveyor.

The application of a vision system to an assembly robot increases its potential, for example, if a component approaches the robot in the wrong orientation, the robot can re-orientate the part and continue with the assembly task. *Unimation* have designed a family of robots specifically for assembly tasks — their *PUMA* (Programmable Universal Machine for Assembly) robots have reached a high level of sophistication and are installed in factories world-wide.

Welding

Welding is a process of joining metals by fusing them together, unlike soldering or brazing where joints are made by adhesion of the surfaces with a joining medium.

Welding is a common application area for robots, and automatic welding has been carried out for many years. The motor industry benefitted enormously from auto-welding procedures, as the many hundreds of welds required to build a car body could be carried out quickly and systematically. Two main types of welding exist: *spot welding* and *arc welding*.

Spot Welding

Spot welding is a system of joining metal in small localized areas or spots, by passing a large electric current at low voltage through the metal at each point to be welded.

A typical spot welding sequence would be as follows:

1. Hold the two pieces of metal together by exerting a force on the electrodes.
2. Activate a current to generate heat in the vicinity of the electrodes.
3. Turn off the current and keep the electrodes under pressure until the metal cools sufficiently for the weld to hold.
4. Release electrode pressure and move to next position.

This sequence of operations is shown in Fig. 9.13.

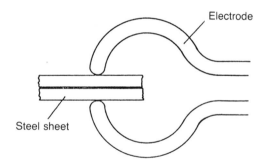

Fig. 9.13. Stages in spot welding: (a) Hold sheets together.

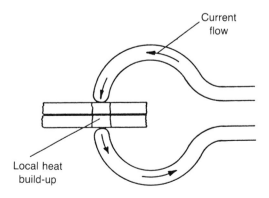

Fig. 9.13. Stages in spot welding: (b) Pass current.

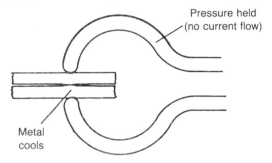

Fig. 9.13. Stages in spot welding: (c) Maintain pressure while metal cools.

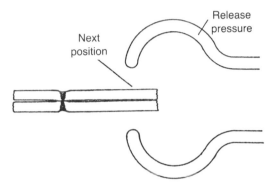

Fig. 9.13. Stages in spot welding: (d) Release pressure and move to next position.

Robot spot welding of car bodies is now commonplace in many car assembly plants. The number of panels to be welded depends upon the vehicle design and the manufacturer, but typically the seams to be welded include: roof panel to side panel; floor pan to side panel; wheel arch to rear quarter panel; and bulkhead to side panel.

An example of an automatic robot welding system is in the *Ford Motor Co.* car assembly plant in Valencia, where a team of eight robots simultaneously spot weld a car body in 132 seconds (this time includes automatic step-wise movement of the body through the work area). Maximum cycle time for any one robot is about 60 seconds, with the fastest robot carrying out 54 welding operations in that time.

Arc Welding

Arc welding techniques are used when a gas-tight seal is required along a continuous path, or where the metal to be joined is too thick to be spot welded.

Arc welding is a fusion technique where the heat required to fuse the metal together is provided by an electric arc. When the arc is struck, the local temperature rises rapidly to about 3600°C, causing a pool of molten metal to form. An electrode in the form of wire is fed into the weld to replace the metal consumed by the welding process. A common arc welding application is the welding of seams on containers or similar components. The robot arm, with welding torch attached, follows a seam providing a continuous weld. The more sophisticated versions with vision systems enable seams to be followed with a greater degree of accuracy. Figure 9.14 shows a robot performing an arc welding operation.

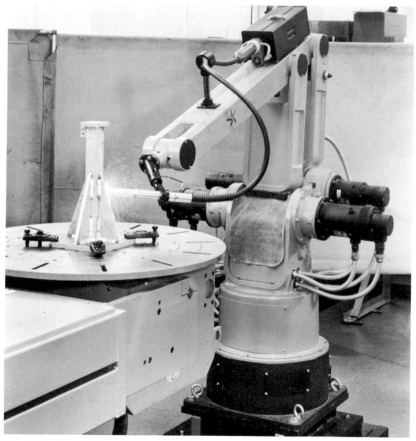

Fig. 9.14. Arc welding robot in operation (courtesy GEC Robot Systems Limited).

Glass Handling

In the majority of assembly-orientated tasks, robots are usually inferior to human beings, as they lack manual dexterity and the freedom of movement provided by the human arms, hands and fingers. In glass-handling, however, this is not the case, as the robot can be equipped with end-effectors which in some ways make it superior to the human operator.

When a human operator picks up a sheet of glass it has to be held around the edges and the length of stretch of the arms limits the width of glass that can be handled by one person. The robot does not use this technique — instead it uses vacuum cups which press against the surface of the glass sheet. Air is evacuated from the cup allowing the sheet to be picked up and moved around. Extremely large pieces of glass can be handled by increasing the number of vacuum cups. Release of the vacuum enables the glass sheet to be put down in any desired location.

There are many glass-handling applications in manufacturing industry. One typical example, however, is the automatic fitment of car windscreens. The system here described is one developed by the *V.S. Technology group* and is a system designed to prepare and fit windscreens and rear windows to car bodies on a moving track. One such system is currently working on the Austin Rover Montego assembly line.

Following preparation of the glass, including priming and adhesive application, the screen is ready for fitment to the car. On the final assembly track the car body is advanced by a special 'lift and carry' transfer to the fitting position.

A robot then collects the glass from a conveyor and positions it approximately 30 mm in front of the car body aperture. The glass is held by vacuum cups which can be adjusted in X and Y axis and for yaw by stepper drive motors. The glass is carried by the robot on a vision-controlled alignment fixture which incorporates four linescan TV cameras together with miniature arc lamps. Each camera observes the relative position of the glass to the edge of the aperture (see Fig. 9.15).

Under independent computer control the stepper drive motors manipulate the glass until it is acceptably positioned relative to the body aperture. Pneumatic pressure is then used to insert the glass. Figure 9.16 shows the intelligent robot inserting the glass into the aperture.

While one robot fits the windscreen, another robot fits the rear window. After assembly the car is replaced on the moving conveyor track in its former position.

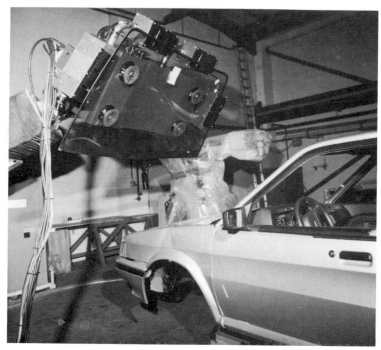

Fig. 9.15. Positioning of windscreen glass relative to car body aperture (courtesy V.S. Technology Group).

Fig. 9.16. Robot inserting glass screen into aperture (courtesy V.S. Technology Group)

Paint Spraying

Spray painting is a task which benefits enormously from the application of robotics. The paint spraying environment is one of the worst that humans have to endure and trying to maintain a dirt-free, temperature controlled area is a difficult task for manufacturing management as a constant supply of fresh-air is required by the operator. Robots can replace the human operator in many spraying tasks giving huge cost savings and many other benefits to industry.

The best suited robots to painting tasks are the *playback* (continuous path) type, in which the robot is taught the art of spray painting by an operator. The operator leads the robot through its painting sequence, then the robot can replay the action when required. Extremely complex components can be sprayed in this way. The majority of painting robots are statically mounted in a booth or air-conditioned area and the work passes on a conveyor or track. The early types of painting robot were simple in design and lacked intelligence. The drawback was that they could not recognize the absence or presence of a component. If, for example, a component was missing from the conveyor, approaching the spray area, the robot would meticulously and lovingly spray fresh air. Modern systems incorporate intelligent devices which detect the presence of components. The fact that high viscosity substances can be sprayed by robots has increased the application potential of spraying robots. Figure 9.17 shows a robot spraying underseal onto the inner wing of a car body. The main benefits of using robots in paint spraying include:

No need for humans to work in hostile environments.
Reduced air circulation requirements.
Less dirt in the atmosphere.
Consistent quality levels.
Reduced material and labour cost.

Machine Tool Loading

The loading of machine tools by robots is gaining increasing importance as companies strive to save shop floor labour costs. The growing installation of flexible manufacturing systems has lead to rapid developments in the techniques employed by machine loading robots. The introduction of DNC and allied manufacturing technology has also increased these developments. Figure 9.18 shows a robot loading a component in a machine tool.

Fig. 9.17. Robot spraying underseal onto inner wing of car body (courtesy GEC Robot Systems Limited).

Fig. 9.18. Robot loading workpiece into a machine tool (courtesy Dainichi-Sykes Robotics Limited).

A large number of manufactured components require several different machining operations performed upon them. If a component travels down a transfer line from one end of the factory to the other, it is desirable to keep it moving down the line, passing it from one operation to the next. A robot system can be set up to feed the component into the machine, let the machine perform the operation, then have the robot remove the component from the machine and place it on a conveyor in readiness for the next operation.

There are certain drawbacks with the one-robot, one-machine layout in that the robot is idle during the time it takes for the machine to perform the machining. To overcome this problem the robot can be in the centre of a work area with the machines clustered around it. The one robot can service a number of machines giving improved robot utilization (see Fig. 7.10).

Robots and CNC machines are complementary in terms of their control systems. By effectively using input/output channels on a controller it is possible to obtain sophisticated control over several items of peripheral equipment, making it possible to integrate robots within an overall flexible manufacturing system.

9.7 ARTIFICIAL INTELLIGENCE

To discuss artificial intelligence (AI) in great detail is beyond the scope of this text book but it should be pointed out that increasing the intelligence of second-generation robots is one factor which will greatly influence the manufacturing effectiveness of industry in the next decade.

Artificial intelligence really means artificial behaviour, by computer programs and robots. AI is the science of making machines perform tasks that would require intelligence if done by humans. The types of activity involved would be: exploration, pattern recognition, learning, and problem solving. At present, artificial intelligence programs accomplish tasks differently from a human being, sometimes better and often quicker. Since digital devices are used for information-processing, all human knowledge must be expressed in terms of logical relations, which means the computer model is still very distant from the human brain.

The main difficulty with AI comes in defining the parameters that the robot is to react to, and to specify the action and quality of response. Robots, in their present form, use conventional feedback techniques in a rather sophisticated manner, but they are still only mechanical handlers. Robots lack the perception associated with human beings which includes all the primary senses and a variety of other mechanisms by which humans sample their environment.

However, with continuing research, the likely emergence of intelligent, versatile robots will have consequences not only for industry but for medicine, education, the armed forces and many more activities.

Chapter 10

The Benefits and Limitations of CAD/CAM

10.1 INTRODUCTION

The benefits of specific areas of CAD/CAM have already been detailed in respective chapters. The purpose of this chapter is to summarize those benefits and present an overall picture of the advantages to be gained from using computer-aided design or manufacturing systems.

There are many benefits of computer-aided design and manufacture. Some are tangible and can easily be monitored, and some are intangible and are not so easily measured, as they are spin-off benefits as a result of using CAD/CAM systems.

The benefits of CAD/CAM can be identified under several headings.

10.2 BENEFITS

Productivity

CAD/CAM as a technology has shown great promise for improvement in productivity. The drawing office is just one area where productivity has been particularly increased due to the introduction of CAD.

There are a number of reasons for this:

(i) High-speed plotters can produce a detailed working drawing in a matter of a few minutes compared with the number of hours it would take a draughtsman to complete.

(ii) As standard symbols and procedures can be recalled from computer memory, the drawing can be constructed with ease in a relatively short time. CAD, therefore, eliminates many of the repetitive drawing tasks. New drawings can also be created by recalling a drawing of a similar nature and carrying out a few amendments.

Draughting systems enable the user to take a number of *short cuts* during the drawing process by the provision of such facilities as mirroring, repetition commands and so on. The actual increase in productivity depends upon the type of drawing produced.

However, the more complex the drawing, in terms of dimensions and repetitions, the greater the increase in productivity.

CAD systems, generally, are a great deal faster than traditional manual systems. It is possible, therefore, with a CAD system to produce a finished set of component drawings and associated quotes and reports in a relatively short time. Shorter lead-times in the design process means shorter elapsed time between receipt of an order and delivery of packaged product.

Costs

The introduction of any new technology brings with it a reduction in the number of people required to operate the system. CAD/CAM technology is no different and dramatic savings in labour costs can be realized with the introduction of CAD/CAM.

Cost savings can also be made in other ways. With a draughting system, for example, whilst the initial capital costs are higher than for a manual system there can be an overall saving in costs over a given time period.

There is less requirement for drawing materials as only the finished drawing will be plotted out. Also, the draughtsman will be able to complete more drawings in a given time using CAD in place of a manual system.

Cost savings on the shop floor can be brought about by a reduction in scrap, as components can repeatedly be produced to consistent quality levels using CNC machine tools. Additionally, machine and tool prove-outs can be carried out on a graphics screen, thus saving valuable time in setting up machines. Computer prove-out also eliminates the possibility of expensive machine clashes.

As geometric modellers and finite elements can produce software models of components, there is a reduced requirement for expensive and elaborate prototypes.

Accuracy

It is characteristic of CAD/CAM systems that the equipment used lends itself to accuracy of designed and manufactured products. Drawings are created and dimensioned to within fine limits and the fact that a plotter is used for production of the drawings means that good line quality and precision lettering are virtually guaranteed.

Tool wear compensation features on CNC machine tools means that components are accurately machined.

With CAD/CAM there is a high level of dimensional control, far in excess of what can be achieved manually. With 32-bit processors and double precision arithemetic the computer system can perform calculations to great accuracy in a very short space of time. Accuracy in the design phase also pays dividends in the manufacturing phase.

Improved Designs

The features in a computer-aided design system help to consolidate the design process into a logical pattern of working. Instead of having a continuous to and fro exchange between design and analysis groups, the designer can perform the analysis of design at the workstation. This improves the concentration of designers as they interact with their design in real-time.

The effective use of CAD leads to a better designed product. Improvements in design come from a number of factors:

○ Modelling systems play a major part in the improvement as components can be checked for interference etc., and good design solutions can be reached in a relatively short time. Volume and mass property details are readily available from models, giving the designer vital accurate information about the design. Mechanism motion analysis can also be carried out to aid the design of mechanisms and linkages, helping to specify the size and shape of housing required.

○ Finite element methods are particularly useful for improving product design by solving engineering problems concerned with stress, vibration and heat transfer.

○ Engineering components which have awkward profiles can be analyzed and key problem areas identified. The use of finite elements is particularly relevant in the aerospace industries where structures possessing high strength and low mass are required.

○ Since amendments in preliminary designs are easy to effect and analyze with a CAD set-up, more design alternatives can be assessed in the available development time. Consequently, this results in the emergence of *better* designs within a given period.

Product Quality

High quality runs right across the CAD/CAM spectrum, from the initial design to the finished packaged product. Having designed the component to high standards, the manufacturing systems can produce components to those design specifications. Computerized pre-process, in-process and post-process monitoring ensures that the raw materials are of the right quality and type, that the work on the shop floor is being carried out to a given specification, and that the finished product is meticulously examined and tested.

Integrated CAD/CAM

An integrated CAD/CAM system means that a central database is used to hold the design and manufacturing information. This means that the information is *consistent* throughout the company and has many benefits apart from the specific design and manufacturing ones. Once the product has been designed, much manufacturing information can be generated automatically, saving part-programming effort and so on.

Stock control and purchasing departments can have access to material and parts lists automatically created in the CAD system. Production engineering departments can plan ahead and schedule the loading of the workshop well in advance, as tooling and machine requirements will be known for any given project.

Management can have timely information about any product and can monitor the exact state of work in progress. Better forecasting and budgeting can be determined using up-to-the-minute information.

Machine Tool Aspects

There are many advantages associated with the use of numerically controlled machine tools. NC, for example, can increase the proportion of time that the machine tool is engaged in the actual process, by having fewer set-ups, reduced set-up times, reduced workpiece handling time, auto-tool changers, and so on. Manufacturing lead-times can be reduced as the jobs can be set up quickly. Greater manufacturing flexibility can be introduced as component design changes can be readily carried out. Quality control is improved by using NC methods as the testing, inspection and quality checks can be automated.

Computer-assisted and graphical part-programming of machine tools also brings with it the following benefits:

○ The part-programmer can construct a program in a much shorter time compared with conventional methods of part-programming. The use of sub-routines leads to a more productive programmer.

○ Programming for complex parts is less difficult. Lead-times are reduced and fewer prove-outs are required, which means the machine tools are tied up for shorter periods.

o The use of Direct Numerical Control (DNC) has certain advantages over NC and CNC systems. For example, DNC eliminates the need for punched and magnetic tape for machine control. DNC provides the opportunity to perform computational functions more effectively than NC, as these functions are implemented with software rather than hard-wired devices, thus increasing programming flexibility. Part-programs can be stored more conveniently as computer files than in punched tape form. The DNC facility would also play a large part in a totally automated factory.

Robotic Aspects

The major benefit of using robots in manufacturing is an increase in productivity. Productivity is gained in both programming the actions of the robot and in designing the workplace for optimum performance. The programming of robots can be a lengthy, skilful task, their application therefore will, in the main, be for high-volume, mass production.

Robots can be put to work in hazardous and unpleasant environments and will perform repetitive tasks without question — thereby improving the health and safety of workers in industry.

Robots are particularly good at palletizing and loading machine tools and are an *essential* element in a flexible manufacturing system.

Standards and Procedures

In an integrated CAD/CAM system the single database is common to all work-stations and machines. Consequently, the system provides a standard for all design, draughting and manufacturing procedures. Drawings and documents will all contain information produced to a consistent, company-defined format.

Original drawings, model descriptions, and documents are stored in the central database. This makes them more accessible than documents stored in drawing and filing cabinets. Information can be *quickly* and *easily* amended then written back to files in its updated form.

Market Image

One of the more intangible benefits of using CAD/CAM is the improved market image of the company.

Quotations and design calculations can be despatched to customers very quickly. Amendments to drawings can be turned round within 24 hours. The drawings and documents sent to customers will be of professional high quality. These factors alone may lead to an increase in sales, as a customer who receives the professional drawing will assume that the finished product will be of the same high quality.

The inclusion of coloured 3-dimensional views of components in sales brochures gives the company a high-tech image.

10.3   LIMITATIONS

Although Computer-Aided Design and Manufacturing systems bring many benefits to industry, there are also some disadvantages to consider.

Social Implications

The introduction of any new technology usually results in a reduction in the number of workers. CAD/CAM is not unique in this respect and as a direct result of the increased productivity benefits of this technology, fewer employees are required to operate design and manufacturing departments. CAD/CAM systems have already been responsible for a number of skilled draughtsmen and production engineers to be out of work, and this upward trend in unemployment is likely to continue into the future.

The introduction of CAD/CAM systems will, in the long term, have a dramatic effect on the structure of companies. It will be met with resistance in certain areas and there may be confrontation between computer-biased engineers and those with traditional methods of working.

The effective use of graphics display terminals requires controlled, subdued, indirect lighting, and windows need to have blinds fitted to reduce glare and reflection. As a result, users operating a CAD system for long periods can suffer from eye strain. Stress can also be caused by the high intensity of working, especially with some of the fast-acting interactive design systems. The attention to ergonomics for a CAD environment is, therefore, more acute than for conventional draughting and design offices.

Costs

Hardware and software for CAD/CAM can be very expensive and a company must justify the acquisition of computer-aided facilities.

Investment costs for draughting, modelling or NC equipment can be a massive drain on a company's capital. The investment programme must, therefore, be considered in the light of the potential benefits of CAD/CAM systems.

High technology CAD/CAM equipment will require maintenance at specified intervals and most system manufacturers will provide a maintenance contract scheme, although these can be very expensive necessities.

The use of a computer-aided design or manufacturing system will require the user to have particular skills. For example, a draughtsman using a draughting system, as well as requiring knowledge of traditional drawing concepts, will require different skills in order to get the most out of the system. To send potential operators on a training course, as well as losing valuable man hours, also incurs a cost to the company.

Chapter 11

CAD/CAM — The Future

11.1 INTRODUCTION

Since the early 1970s, computer-aided design and manufacturing systems have provided industry with hope and excitement about the prospect of this relatively new technology. CAD/CAM has responded to the needs of industry by the provision of sophisticated design tools, computer-controlled machine tools, advanced robotic systems and so on.

To predict future changes in technology is a notoriously difficult task. The purpose of this chapter is to provide sensible suggestions regarding the developments in CAD/CAM which are likely to take place during the next decade. Various aspects concerning the future developments in CAD/CAM will be discussed separately.

11.2 FUTURE ASPECTS

Trends

Recent surveys predict that CAD/CAM systems will be introduced into manufacturing industry on an increasing scale over the next two decades, leading to computerized automation of shop floor processes with subsequent changes in working environments and social conditions. Industry will see a convergence of design and manufacturing activities into centralized *integrated* CAD/CAM systems and there will be an increased reliance on computer-based facilities in areas that were traditionally labour intensive.

These trends are expected to continue and companies will be forced to adopt new technology in order to remain competitive.

Hardware and Software

The early CAD/CAM systems could be operated only on large expensive mainframe computers. However, the past 5 to 10 years has seen an increase in the use of minicomputers for CAD/CAM applications.

240

System suppliers recognized the need to cater for users of mini-computers in CAD/CAM and many of the major computer manufacturers introduced a powerful range of mini-computers, called *super-minis*. As the price of 32-bit machines falls, more and more companies can cost-justify CAD/CAM equipment. Graphics display terminals are becoming increasingly sophisticated and more and more local intelligence is being incorporated into them. A distributed CAD system with intelligent work-stations will be a serious competitor for current CAD systems. Other peripherals will also be more effective as a result of local intelligence. A plotter, for example, with this facility could be capable of plotting complicated shapes based on relatively simple instructions given to it from the mini-computer.

The density of information that can be stored on a magnetic disk is increasing at a rapid rate and the trend looks to be towards the use of enclosed hard disks like the Winchester. Other mass storage media looks set to play a part in CAD/CAM systems, such as *laser disk* or *bubble memories*.

There will be parallel development of software to complement the hardware advances. The development of high-level languages will reduce implementation costs for engineering software, and the use of database management systems, which are limited at present (although widely used in business applications) is likely to increase.

As an incentive to industry to implement CAD/CAM, the Department of Industry is providing £12 million during 1983–85 to help companies in the UK accelerate the introduction of relevant hardware and software.

Manufacturing activities will become more sophisticated as voice recognition and vision systems develop. Voice input to an NC machine tool for part-programming is already a reality and robot technology will exploit sound and vision systems as vision-controlled tactile and proximity sensors will become standard elements of manufacturing robots. These developments will almost certainly lead to an increase in the use of robots in more diverse manufacturing environments.

The Role of the Micro

Until quite recently there were very few CAD/CAM systems operating on microcomputers because of the many limitations attached to this category of computer.

For example, the lack of processing power of the 8-bit microprocessor which found difficulty in executing arithmetic operations required for the effective manipulation of design and manufacturing data. The absence of high-level scientific languages for these machines made it difficult for engineers to develop analytical software. Finally, the mass storage capability of microcomputers was based on floppy disk systems which have a physical limit on storage size.

However, the trend is changing and many of the limitations of microcomputers have been overcome. Modern microcomputers now employ 16-bit microprocessors which make the processing power comparable with some mini-computers. More sophisticated disk-based operating systems have been developed and high-level languages such as PASCAL and FORTRAN can now be compiled and run on a micro.

Microcomputer systems have taken advantage of the development in mass storage technology and can now have extensive data store capabilities as a result of enclosed hard-disk systems like the Winchester.

Due to their relatively low cost, microcomputers are an attractive proposition for CAD/CAM buyers and more and more micro-based CAD/CAM systems are entering the market.

Even the traditional mainframe-orientated activities, like finite element analysis, can now be carried out on microcomputers, and some manufacturers have recently introduced finite element software to run on desk-top micros.

If this overall trend continues the result will be to bring CAD/CAM technology to a wider range of engineering and related companies.

Integration

The application of new technology to the individual aspects of design and manufacture is now well understood and is proving beneficial to many companies. However, the total integration of these two areas into one common activity is still under development. In order to gain maximum benefit from a total system it is *essential* that this integration takes place. Integrating design and manufacture in this way will certainly change the way in which manufacturing companies operate. With increasingly sophisticated communication methods there will be the opportunity for a customer in one country to pass a component specification, via satellite, to a manufacturing computer in another country and have the component manufactured in a short space of time. It is not unreasonable to foresee a designer or engineer designing and specifying a machined component at a work-station and then letting the system make the product without the need for further manual intervention.

CAD/CAM system manufacturers are marketing products which are no longer designed to stand-alone. The majority of the newly developed systems have the ability to link with other aspects of design or manufacture. As this trend continues there will be a gradual swing away from the traditional separation of design and manufacturing activities and instead towards total integration.

Education

One of the greatest impacts of CAD will be that it will demand the formalization of design procedures and the thinking through of the relation of design to manufacture in conjunction with the manufacturing engineer. The ready availability of databases and software to calculate the effects of design changes will reduce the dependence of the designer on the specialist. CAM, together with its associated discipline of design for economic manufacture, will make new demands on the designer and change the criteria used in evaluation and formulating a new design.

CAD/CAM itself will strengthen the links between design and manufacture: the design department will simply no longer be able to be treated or to act as a separate part of the manufacturing process. The implications of this will need to be reflected in the education and training of designers.

Students and managers alike will need to be educated and trained in the principles and use of CAD/CAM. Already, educational institutions have established courses in CAD/CAM at diploma, graduate and post-graduate level. Many of the engineering courses and some computing courses also have a CAD/CAM module in their syllabus. Vast amounts of money are being spent in many countries on CAD/CAM hardware and software to support current and future courses.

Managers, too, can gain education and user-experience in CAD/CAM and to provide an incentive to managers to update themselves in this new technology the UK Government Department of Trade and Industry is providing information, training and consultancy through their CAD/CAM management awareness programme. Here, executives can go along to a venue and seek advice on CAD/CAM from specialists in certain fields. Practical experience centres also provide an opportunity to gain hands-on experience in real-life working situations. Under the DTI's scheme, more than 2450 company executives were introduced to new CAD/CAM technology over a two-year period.

1st January 1985 marked the start of a new awareness campaign in the UK, linked to a six-figure contract between the Department of Trade and Industry and the Institution of Mechanical Engineers, in which the I.Mech.E will administer a programme of visits to firms demonstrating the uses of advanced manufacturing technology.

During 1985 the Demonstration Firms Scheme will aim to take 400 visitors to 100 firms to see CAD, CAM, CAPM, FMS and robotics in an industrial environment. Such visits are seen as unique opportunities to find out not only the benefits, but also the problems of installing AMT.

The involvement of governments underlines the growing importance and the need for an increased awareness of computer-aided design and manufacture throughout the world.

Appendix 1

CAD/CAM Systems

This appendix is designed to provide a guide to some of the major international system suppliers.

CAD/CAM system suppliers can be classified into the following categories:

- graphics display terminal manufacturers, that offer CAD/CAM software (usually stand-alone systems, intended for specific applications);
- mainframe and minicomputer manufacturers who have developed their own software (to run on general purpose computers or as a dedicated system);
- software houses who either provide their own software or who also market third party software;
- companies offering Turnkey CAD/CAM systems. A turnkey is a package of compatible hardware and software (that may or may not be from the same manufacturer) that is sold as a complete ready to use system.

A 'quick reference' code is provided to show the areas of specialization for each company.

The list of applications of each system serves to underline the wide scope for the use of CAD/CAM, from PCB design, through architectural drawing to robot cell simulation. Because of the overlap between CAD/CAM products and the companies which market them, the majority of entries in this appendix are turnkey systems. No attempt is made to make comparisons between or discuss the relative merits of system suppliers.

The developing nature of CAE technology is such that product changes are made frequently, as companies review and enhance their systems. Therefore, specifications are only correct at the time of going to press, and where prices are given (correct for 1985) they are only indicators to the cost of the system.

The name of each organization is set in bold type, followed by the name of its principal product, where appropriate. Descriptive classifications are then given according to the following system of abbreviations:

Key: 2DD   2D Draughting
3DD   3D Draughting
3DM   3D Modelling
FEA   Finite Element Analysis
M   Manufacturing (including NC and Robotics)
T   Turnkey

AM Admel   EASYDRAF^2 2DD

A 2D drafting system based on Hewlett Packard computers. A graphics tablet or mouse drives a cursor and dynamically changing on-screen menu. Features of the system include pre- and post-processing the database via ASCII file format, background plotting, file management and sorting utility. Price from £10000 for 9816S based system. Applications include: mechanical and electrical, schematic, architectural, construction, plant and office layout draughting.

Applicon   BRAVO   2DD/3DD/3DM/M/T

The Bravo family of turnkey systems is based upon DEC VAX 32-bit computers and Applicon workstations run under the DEC VAX VMS operating system. The Applicon editor is a structured menu-driven graphics system for creating, editing and maintaining engineering design and drawing databases. Hierarchical on-screen menus with user-selectable prompting levels, combined with Applicons tablet symbol recognition, make the system easy to use. Other applications software packages integrated within the system include solids modelling, surface modelling, finite element analysis, bills of materials, flat pattern development, PCB design, NC, etc. Price from $155000 for VAX 11/730 based system.

Applied Research of Cambridge   ARCAD   2DD/3DD/T

ARC has long experience in the practical use of computer graphics and specializes in large interactive computer systems using advanced software techniques. ARC produce GDS (General Drafting System), one of the most powerful and versatile drafting systems available. The Arcad software family consists of GDS as core graphics module, supported by optional modules. GDS has automatic scheduling, 3D visualization and user programmability. Capabilities can be enhanced by adding the superview system (SVS) for creating solid colour 3D views. Arcad modules address various application areas including: architectural, civil, structural and building services engineering as well as mechanical and electrical engineering and mapping. Prices from £60000 for Prime 2250 based system.

Autodesk   AUTOCAD   2DD

AUTOCAD is a 2D micro-computer based program available to run on a wide range of micros including: Apricot, Columbia, Compaq, DEC Rainbow etc. Drawings are created and edited on-screen with a digitizer or mouse. Drawings are produced on layers with up to 255 colours. Advanced drafting extension 2 adds object snap: dragging for circle, arc, shape, insert, move, copy, change commands; isometric and rotated grids; and attributes, with database extraction facilities which can be linked directly to dBase II and other database packages. Price from £6400 for IBM PC system. Software only available at £1600.

Computer Aided Design Centre   DIAD/GNC   2DD/M/T

DIAD is a 2D drafting system running on the Apollo Domain workstation for £42650. CAD Centre also have a Graphical Numerical Control system (GNC), and a 3D surface modeller.

Cadlink   CIMSTATION   2DD/3DD/3DM/M/T

The building blocks of Cadlink's Cim system are stand-alone 32-bit engineering workstations. Each Cimstation comprises a graphics display monitor, a detachable keyboard, a mouse, and a cabinet housing a Motorola 68000 CPU with 2.5 Mbyte DRAM, 30 Mbyte Winchester disc drive and 1Mbyte floppy disc unit. Each Cimstation can be loaded with CIMCAD for 2D and 3D drafting and design, CIMSURF for 3D surface modelling and 5-axis NC programming, CIMCAM for 2.5-axis NC programming and CIMSHELL the Cadlink user interface. Prices start at £55000.

Cascade Graphics Development   CASCADE X   2DD

The Cascade system is the result of joint effort by an international engineering company which has used computer-aided draughting systems for over twelve years and a California computer graphics company. This combination of experienced draughtsmen and advanced graphics programmers set out to produce a CAD system which would be compatible with the needs of all draughting offices. Cascade X is a monochrome or colour 2D draughting system with dynamic pan, zoom, move, copy plus lines, arcs, points, fillets, line trimming, auto-dimensioning, cross-hatching, rotate, mirror, parts listing, and parameterization language. Price from £13000. CGD also sell Cascade 1 on an Apple II for £2500 and on an IBM PC for £4000.

Cambridge Interactive Systems Limited   MEDUSA   2DD/3DD/3DM/M/T

This company developed MEDUSA, a highly successful package for engineering design. Medusa is an integrated modular family of design, drawing and manufacturing aids, with a common database allowing a very high degree of flexibility in any particular application. CIS have recently introduced a Sheet Metal Design system. A choice of workstations can be employed including Tektronix and Westward systems. The Medusa software can be implemented on a range of hardware including: Prime 50 range of 32-bit machines and DEC VAX 32-bit computers. Applications include: 2D and 3D design, 3D solid modelling, Structural engineering, piping and plant layout, architectural design, schematics, and logic diagrams. CIS is now a Computervision company.

Computervision   CDS 4000   2DD/3DD/3DM/M/T

Computervision is probably the world leader so far as the number of systems installed is concerned. The CDS 4000 is CV's mainstream product based on their own hardware. Modules include: ASD for surface design, DRAFTEZE for 2D drafting, SOLIDESIGN for 3D solid modelling and ROBOGRAPHIX for robot cell simulation. A 16-seat system costs $80000 per workstation. CDS 3000 is a sun workstation-based system with icon-driven graphics. Applications software includes 2D drafting, schematic capture for logic design, space planning and technical publications. Price from $38500. CDS 5000 is an IBM 4300-based system running group technology software, frontended by a CDS 4000 or Designer-VX. Price from $490000.

Counting House   ITS   2DD/3DD/3DM/M/T

Established in 1976, Counting House computer systems achieved rapid growth through sales of its own integrated business system, based on Prime computers, to engineering companies. At the same time the company became the leading supplier of the graphic NC part-programming system, GNC, enhancing the system with its own software and with a wide range of post-processors. Counting House now offer 2.5D, and 3D design, machining programs and CAD/CAM interfaces. ITS is a Prime-based modular system comprising: drafting/design, drawing/data management, 3D solid modelling, NC programming, manufacturing control and financial accounting. Price for Prime 2250-based system from £75000.

Delta Computer Aided Engineering Limited   DELTACAM
            2DD/3DD/3DM/FEA/M/T

Delta CAE Ltd was originally formed, in the early seventies, as a Technical Services Unit to advise and assist Delta group companies in introducing NC machine tools into their tool rooms for die and mould manufacture. Today, Delta CAE offers an advanced, integrated computer-aided design, draughting and manufacturing system on a modular basis either for in-house or use as a bureau service. Deltacam comprises three main modules; for 2D drafting, 2.5D NC programming, and 3D modelling and machining. The 2D drafting system is suitable for all engineering drawing applications, with facilities for storing standard parts, families of parts and producing automatic parts lists. GNC is a 2.5D NC programming system for milling, turning, flame cutting, wire erosion, nibbling and punch pressing machines, which is interfaced with DOGS (DOGS is a product of Pafec Limited). DUCT is a 3D surface modelling system for the design and manufacture of complex shapes such as those found in plastic mouldings and forgings. The system is suitable for all stages of product development from conceptual design, detailed design to model making and tool making. Deltacam runs on DEC VAX, Prime and Apollo computers. Deltacam provides additional sytems for process planning, manufacturing control, shop floor data collection and database management. Price £40000.

Denford Machine Tools Limited   2DD/M/T

Denford Machine Tools produce a range of CNC and DNC training equipment, including hardware and software, and in 1984 won an Industrial Achievement Award. Denford's have recently launched a 2D draughting system for BBC/Acorn users. It incorporates features often associated with larger expensive CAD systems. The ORAC is a CNC bench lathe which includes toolpath graphics and a wide range of software packages for different microcomputers. The STARTURN DNC desktop training system links a mini lathe to a BBC/Acorn computer developed for classroom situations in schools, colleges or training centres. Denford also supply EASIMILL 3, a CNC turret mill, EASITURN 3, a CNC lathe, and an ORBIT training robot.

Engineering Computer Services Ltd   GRAFTEK   2DD/3DM/FEA/M/T

GRAFTEK is a second generation integrated 3D system with modules for drafting, solid modelling, 5-axis NC, finite element modelling, and plastic flow analysis, designed to run on 32-bit super minicomputers. The software AGILE is a customizing language designed to run on HP9000, VAX and gould SEL

32/xx computers. Price from £90000. ECS also market PULSAR, an HP200 based NC tape prep system for £10000: FESDEC, a desktop finite element system for £10000: and CAD200, a 2D draughting system running on HP200 series computers for £25000.

Ferranti Infographics   CAM-X   2DD/3DD/3DM/FEA/M/T

CAM-X is a DEC VAX based CAE system with integrated and flexible software. The amount of functionality can be sized to match the user requirements for applications in the design and manufacturing engineering industry. CAM-X facilitates a wide range of software including: 2D design and drafting, 3D solids modelling, finite element mesh generation, NC programming, engineering records management (ERMS) and graphics library for user extensions (GLUE).

Hewlett Packard   HP-DRAFT   2DD/3DD/T

HP-DRAFT is a versatile 2D/2.5D draughting system that allows you to create drawings ranging from mechanical to architectural. It is best suited for creating new documentation and making drawing revisions. The software can also be applied to tasks such as assembly and proposal drawings, layouts and schematics. HP-DRAFT can be used as a stand-alone product on HP9000 series 200 computers or in a shared resource management network, sharing common peripherals (disk, plotter, printer etc.). HP-DRAFT can also be linked, via HP design link, to HP-FE and HP-NC for complete CAD/CAM. Price from £30000. HP also market EGS/200 for PCB design for £30000.

Integrated Micro Products   MDS   2DD

The Micro Drafting System from Integrated Micro Products and Advanced Computer Graphics is a general purpose 2D drafting package suitable for a wide range of applications in architecture, engineering and the electronic and electrical fields. It is written in Fortran 77 and runs on the IMP-68 multi-user computer under the Xenix (Unix system III) and Idris (Whitesmiths Ltd Unix) operating systems.

Intergraph   INTERGRAPH   2DD/3DD/3DM/FEA/M/T

Intergraph computer graphics systems are based on the DEC VAX range of 32-bit computers, enhanced by special Intergraph-designed hardware accelerators to provide the speed of response required for interactive graphics. Intergraph systems are supplied with two interrelated software packages as standard: IGDS (interactive graphics design software) which provides a set of general-purpose 2D/3D design and draughting tools: and DMRS (data management and retrieval system) which is a DBMS for handling all non-graphic data associated with the graphics. Software packages include: solids modelling, surface modelling, kinematics, NC programming, and nesting. Together with electrical applications such as autorouting, autoplace, autotest etc. Plus several cartographic and utilities mapping and exploration packages. Price from £110000.

International Research and Development Co Ltd   CADBIRD II   2DD

The CADBIRD system is an integration of hardware and software that is versatile, inexpensive and simple to operate. CADBIRD II is operated by simple keyboard commands and covers draughting needs such as auto-dimension, fillet, chamfer, erase, symbol, zoom, hatch, mirror image etc. It runs on

DEC PDP11/23 PLUS computer with hard disk store. A Westward raster graphics terminal is used for interactive working. CADBIRD II can be directly linked to NC tape preparation systems, giving control of the complete process from design to manufacture.

Marconi   MICROQUAD   2DD/3DD

MICROQUAD is a printed circuit board CAD system that integrates the design office and the factory. The computer module has portable desktop packaging and consists of a Data General micro ECLIPSE microprocessor with 768 kbytes of main store, Winchester disk drive and an IBM PC compatible floppy disc unit. Features include: data input systems, connectivity check, design rule check, component placement, interactive editing, auto routing, and manufacture and test outputs.

Mc Auto   UNIGRAPHICS   2DD/3DD/3DM/M/T

UNIGRAPHICS is an interactive graphics system with applications in engineering design and analysis, draughting and the generation of NC data for manufacturing. System modules include: GRIP (graphics interactive programming language) which allows the user to automate the functions of Unigraphics, and modules for lathe, mill and 5-axis machining centre programming. UNIGRAPHICS is available on Data General MV and DEC VAX computers. Over 70% of Unigraphics users use the system for NC tape preparation. Price from £70000 for a Data General MV4000 based system. Other products include: UNISOLIDS for solids modelling and MRS for robot system design.

Micro Aided Engineering   2DD/3DD/FEA/M/T

MAEDOS is a 2D drafting system that has mirror, layering, parameterized symbols, parts list from the drawing and 3D visualization. MAECAM is an NC programming system that gives graphical verification of the geometry, tool and tool path to reduce prove out time. MAEDOS can be linked to MAECAM, which can then be linked to the machine tool. There are also various systems covering digitizing, curve fitting, nibbling, post-processors, estimating and costing. MAESIM is an NC verification and simulation system. MAE have recently introduced MAEFIN, a finite element system. All MAE systems use 15 and 19 inch colour terminals and the micros can be upgraded and networked. Price from £12750 for an IBM XT based system.

Norrie Hill   SOURCE 36   2DD/T

SOURCE 36 is a 2D system based on the HP9836 with links to NC. Norrie Hill also market Tsquare on the IBM PC XT and DRAFTY on the HP86. Price starts at £15000.

Pafec   2DD/3DD/3DM/FEA/M/T

This company initially gained recognition for its finite element analysis package but is now internationally famous for its wide range of CAD/CAM products. DOGS is a 2D drafting and design system. The menu contains the options expected of a sophisticated CAD system together with automatic generation of bills of materials/parts lists, parameterized symbols and a drawing management system. The system supports interfaces which include links to NC part-programming systems (DOGS NC, GNC, APT, COMPACT II), mesh generation facilities PIGS (Pafec Interactive Graphics System) for input to the Pafec finite element package, interfaces to DOGS 3D (wireframe) and BOXER (solids modelling) 3D systems. In 1984, Pafec sold more 32-bit CAD/CAM systems than any other vendor in the UK. Prices from £39000 for Apollo-based systems.

Pathtrace Ltd   PATHTRACE/PATHTURN   M/T

Pathtrace Ltd, formed in 1983, has over 150 users of its CNC programming systems in the UK. Based initially on Commodore computers, the latest systems use ACT Sirius 16-bit microcomputer with a second, colour graphics screen. Pathtrace and Pathturn are low-cost graphical computer-aided programming systems. The software comprises geometry definition, tool library (sectioned into categories with 1800 tool capacity), machining specification, text editor, screen and printer output, external communications etc. Price from around £6000 for Apricot system.

Prime   MEDUSA   2DD/3DD/3DM/T

Prime MEDUSA is a CAD system for 2D and 3D design, drafting and documentation. The software also has modules to handle program development, database admin and system interfacing. Model analysis permits the calculation of physical properties of the model, and the system supports colour operation and colour shading of solid models. MEDUSA runs on all Prime 50 series 32-bit super mini computers. Price from £92000. Prime also market PDGS for sculptured surface design at £150000, EDMS for electronics design at £88000, THEMIS for logic simulation at £15000 (software only) and PDMS for process plant design at £196000.

Radan Computational   2DD/3DM/M/T

Radan Computational Ltd was formed in 1976 in order to provide turnkey integrated CAD/CAM systems for the engineering industry. It was founded by members of the staff of the University of Bath. Products include: RADRAFT, a Tektronix 4054 based 2D drafter for £28500, the RADPUNCH system for the preparation of NC programmes for punching and nibbling machines, the RADAN VOLE 3D modeller, and a building modeller called GABLE.

R.H. Symonds Ltd   SYM/CAM   M

SYM/CAM is a sophisticated in-memory tape creation and editing system for CNC machines based on a 16-bit processor running under CPM-86 operating system. Prices start at £1707, or £2072 to include optional monitor.

SIA Computer Services   STRIM 100   2DD/3DD/3DM/FEA/T

STRIM 100 is a 2D and 3D design/modelling system, designed primarily for the mechanical engineering industry. The system consists of three modules: Strim 100 T 3D modelling and design with surface representation, sculptured surfaces, high accuracy and integral 2, 3 and 5 axis machining; Strim 100 finite element mesh generator, with automatic surface meshing, solid and thin shell elements, interactive modification and interface to any FE analysis system: and Strim 100 C 2.5D drafting and NC machining, with rapid generation and manipulation of geometric elements, hatching, machining simulation etc. Price from £50000.

Appendix 1

Shape Data Ltd   ROMULUS   3DM

ROMULUS is a solid geometric modeller. First introduced by Shape Data in 1978 and now substantially enhanced, it places a tool in the hand of the engineer to model a wide variety of component shapes and to perform geometric computations on them. As shape information forms only part of the complete description of a component or assembly, Romulus data structures are designed so that non-geometric information such as material type, surface finish, part numbers, screw thread details and so on may be incorporated into the model. Thus complete product descriptions may be built up in the computer.

Appendix 2

Glossary of CAD/CAM Terms

ACCESS TIME   The time interval between the instant at which information is called for from storage and the information being available for use.

ADAPTIVE CONTROL   A technique for automatically adjusting feeds and/or speeds to an optimum by sensing cutting conditions and acting upon them.

ADD   A basic system command to enter geometry onto a drawing area in memory.

APPLICATIONS SOFTWARE   Software written specifically to solve a set of unique tasks, i.e. finite element software or draughting software.

AMBIGUOUS   A term used for an incompletely described 3-D model — as at least one face in the model will have more than one possible interpretation.

ANSI   American National Standards Institute.

APT   A universal computer-assisted program system for multi-axis contouring programming.

ASCII   American Standard Code for Information Interchange, to define the precise format in which data is presented and received between systems and devices.

ANALOGUE   Electronic circuitry which represents and manipulates data as an infinitely variable signal voltage.

ARCHIVAL STORAGE   A computer peripheral device used to store copies of drawings, models or programs as a back-up to the main memory (see also MASS STORAGE).

ASSOCIATIVITY (of views)   The ability of a CAD system to reflect changes made in one view in every other affected view on the drawing.

ASYNCHRONOUS COMMUNICATION   A type of communication where every character transmitted is preceded and followed by timing pulses.

BAUD   An expresssion denoting the rate in bits per second of an asynchronous interface, including both timing and data bits.

BIT   The smallest unit of data storage in a computer (Binary Digit).

BLENDING   The joining of two or more surfaces so that their co-ordinates form a single continuous surface. No joins or breaks should be present and the surface geometry should be suitable for direct use by NC toolpath programming applications.

BOOLEAN OPERATORS   A set of three relationships which are used to combine 3-D shapes to create solids models (UNITE, DIFFERENCE and INTERSECTION).

BOOTSTRAP A short sequence of instructions which, when entered into the computers programmable memory, will operate a device to load the programmable memory with a larger more sophisticated program (usually the Operating System).

BOUNDARY FILE A set of edges and faces which describe the periphery of a solid model. This is produced as an output from the solid modeller with hidden lines shown dashed or removed in each view.

BUFFER STORE A register used for intermediate storage for information in the transfer sequence between the computer's accumulator and associated registers and a peripheral device.

BUS An electrical highway used for transmitting signals or information between elements.

BYTE A collection of eight data bits used to represent a character or number.

CAD (Computer-Aided Design) The application of a computer to aid the design process.

CADCAM Used to mean Computer-Aided Design, Draughting and Manufacturing using an interactive graphics workstation.

CANNED CYCLE A preset sequence of machine control events initiated by a single command. Such canned cycles are preprogrammed into machine tools.

CARTESIAN CO-ORDINATES Means whereby the position of a point can be defined with reference to a set of axes at right angles to each other.

CAE (Computer-Aided Engineering) The concept of automating all steps involved in bringing a product to a market and enabling all disciplines to access and share the data created by others.

CAM (Computer-Aided Manufacture) The application of a computer to aid the manufacturing process.

CIM (Computer Integrated Manufacturing) The term to describe the total approach to automating all functions involved with the manufacture of a product.

CIRCULAR INTERPOLATION Capability of generating up to 90 degrees of arc using one block of information in a part-program.

CLASH DETECTION The ability of the CAD/CAM system to detect if an object invades the space of another within a model.

CNC (Computer Numerical Control) Using a computer connected to a machine tool, to accept tool path instructions.

CRT (Cathode Ray Tube) An evacuated glass envelope containing a heated electrode which emits a stream of electrons onto the phosphor-coated inner surface of the tube face.

CPU (Central Processing Unit) The central element of any computer system including memory, arithmetic logic unit and control unit.

COMMAND An instruction to the CADCAM system to perform an operation issued using one or more interacting devices.

COMMUNICATION LINK   Electrical connection between workstations and processors or between processors.

CONTOURING   An operation in which simultaneous control of more than one axis is accomplished.

CO-ORDINATE SYSTEM   A convention used whilst working on a 2-D drawing to direct the placement of geometry.

CSG (Constructive Solid Geometry)   A type of generation of solids model.

CURSOR   A symbol which appears on the workstation screen and moves under control of a pointing device, such as a pen, joystick or mouse.

CUTTER PATH   The path which the cutting tool follows as it moves about the workpiece.

DAISY CHAIN   A method of connecting workstations to each other and to the host system. A high-speed serial interface enters the first system which is then linked by a cable to the second, and so on.

DATABASE   A collection of information stored in a centralized area which can be accessed by many applications programs.

DBMS (Database Management System)   Comprehensive software for the efficient creation, storage and retrieval of information, in a manner which suits the particular user.

DEPTH MODULATION   A facility in the hardware of a vector refresh graphics display which automatically reduces the intensity of displayed lines as they go further from the viewers eye.

DIGITIZER   A large version of the graphics tablet, enabling existing drawings to be entered into a CAD/CAM system by pointing with a puck or pen at line end points.

DISK (DISC)   A mass storage device consisting of a flat sheet of metal coated with a ferrous oxide. The disk rotates at speed providing fast access to information.

DISTRIBUTED PROCESSING   The shifting of activities out of a centralized processor into each of a number of connected processors.

DNC (Direct Numerical Control)   Numerical control of machining directly from a computer.

DOUBLE PRECISION   The use of two computer words to represent a number.

DRAGGING   The act of dynamically moving selected parts of a displayed drawing across the screen, under the control of an interaction device.

DRAWING AREA   The area reserved in a computer memory for each workstation to create drawings (conceptually an electronic drawing board where work is carried out prior to actually storing or plotting the drawing).

DRUM PLOTTER   A drawing device having a continuous drum of paper and one or more pens. The paper movement providing one writing axis, the pen movement providing a writing axis at right angles.

DVST (Direct View Storage Tube) A system whereby a steady stream of electrons is used to maintain a sufficient flow of current to illuminate the phosphor coating on the face of a display screen so that the image can be seen for many minutes afterwards.

ELECTROSTATIC PLOTTER A raster output device which deposits carbon particles onto electrostatically charged spots on paper or other media.

FINITE ELEMENT ANALYSIS A technique in which a component is defined as being composed of many interconnected parts (or elements), each of which will have different forces acting on it. By calculating the individual displacement of each element for a given force applied to the component, the overall displacement of the component can be computed. Also used for the solution of thermal and other engineering analyses.

FINITE ELEMENT MODELLING The process of describing a component in terms of element shapes and their locations. This can be done manually or with the aid of mesh generation software.

FILE A collection of data stored in a computer under a given name.

FIRMWARE Software routines stored in hardware, for example, frequently used procedures to rotate a drawing on a graphics screen. Routines usually stored in semi-conductor ROM (read only memory) chips.

FLICKER A phenomenon which occurs when the image displayed on a refreshed graphics display has faded before the beam of electrons can re-draw that part of the image on the screen.

FORTRAN A high-level algebraic procedural computer programming language. FORTRAN (FORmula TRANslation) is used to write a large proportion of the currently available CAD/CAM software.

FUNCTION KEY A key used to execute a command or call up a standard symbol or part from a library file. Can often be user defined.

GRAPHICS TABLET A small low-resolution digitizing board (see DIGITIZER).

GRAPHICS PROCESSOR Special hardware to perform picture image manipulation (e.g. zoom, rotate) in order to relieve the host processor of these tasks.

GREY SCALE A variation in the intensity of the raster pixels on a monochrome graphics display. Used to differentiate lines in a complex drawing and to shade areas.

GRID A matrix of points, with the vertical and horizontal pitch defined by the user, which may be overlayed onto the screen to aid the production of a drawing under construction. Lines and points may SNAP to any corner point on the grid (see also SNAPPING).

HARDWARE A term used to describe all the electromechanical elements of a computer system (e.g. VDU screen, disk, processor).

HIDDEN LINES The edges of an object which would not be visible from a particular viewpoint. 3-D surface modellers and some wireframe modellers are able to remove these lines automatically.

HOST A computer which provides resources to attached workstations and possibly to other systems.

INTEGRATED CIRCUITS (IC) A collection of microscopic electronic circuits created on a slice of silicon or similar material. Many ICs can contain over 100,000 circuits.

INTELLIGENCE A jargon term for the degree of local processing ability within a CAD/CAM device.

JOYSTICK An analogue operated interaction device having a hand operated arm which is moved to produce either two or three axes of output (*x, y* and zoom). A joystick is used to control cursor movement on a graphics display screen.

LAN (Local Area Network) A form of synchronous communication network where the nodes are able to transfer data between themselves at very high speed and without involving a host computer.

LAYERS Discrete levels of information in a computer-generated drawing. Any one or more layers can be viewed together on the screen, with the ability to turn specified layers on or off (toggle). Separate layers are usually reserved for part geometry, construction lines, hatching and so on.

LINEAR INTERPOLATION A function of a control whereby data points are generated between given co-ordinate positions to allow simultaneous movement of two or more axes of motion in a straight line.

MACHINING CENTRE A machine tool capable of drilling, milling, reaming, tapping and boring multiple faces of a part. Machining centres are often provided with a system for automatically changing cutting tools.

MACRO A source language instruction from which many machine language instructions can be generated.

MASS PROPERTIES Information describing the volume, weight, centre of gravity, moments of inertia and products of intertia of an object.

MASS STORAGE A means whereby large volumes of information can be stored on a permanent basis. Usually implies magnetic disk or magnetic tape storage (see also ARCHIVAL STORAGE).

MENU A file of commands, instructions and sub-routines which can be individually addressed using some form of addressing method. A menu is presented in the form of boxes on a card which can be overlayed, or etched on to the surface of a graphics tablet, or alternatively be made to appear on the face of a display screen. Items from the menu are selected by pointing to the boxes with an interaction device such as a puck or pen. A menu simplifies and reduces the amount of user input required for operation of a CAD/CAM system.

MIRROR A command which makes a copy of selected geometry and then mirrors that copy about a given axis.

MODELLING, SOLID The complete geometric description of the space occupied by an object.

MODEM (acronym for MOdulator/DEModulator) A hardware device used for the connection of a computer system and its terminals to a telephone line.

MOUSE   An interaction input device for positioning the screen cursor. The mouse has wheels or a ball on its underside which turn when the mouse is moved over a surface. The action of the rotating wheels or ball sends x, y co-ordinate information to the workstation.

NUMERICAL CONTROL   A method of controlling the actions of a machine by encoding the co-ordinates of their motion, together with associated commands.

NESTING   The act of fitting individual flat parts within a specified area of material in order to achieve optimum usage and minimum waste.

NETWORK   A term to describe the connection of two or more elements of a computer system to a communications interface.

ORIGIN   The point on a draughting area or modelling plane where the x and y co-ordinates are set to zero.

OPERATING SYSTEM   The major suite of programs which control the running of the computer system. The operating system looks after such items as: controlling transfer of data around the system, allocating disk storage, and error recovery.

OVERLAY   A menu card containing command instructions placed on a graphics tablet.

PAN   A command which moves the viewing window around the drawing area, but without altering the window width.

PARALLEL LINK   An electrical connection between workstation and host system, having separate wires for keyboard, tablet, display screen etc.

PARAMETRIC SYMBOL   A part which has been created with one or more variable dimensions, whose actual value may be dictated by the user. Used to construct families of similar components.

PART-PROGRAM   Specific and complete set of data and instructions written in a source language for computer processing or written in machine language for manual programming.

PIXEL   An addressable point on the screen of a raster display.

POCKETTING   A machining operation to remove all the material within a defined boundary.

POST-PROCESSOR   A software routine which converts data which has been analyzed into suitable form for output to a final stage in a CAD/CAM activity. Post-processors are used for such things as converting NC toolpath information into the correct format for use by a machine tool, or for converting finite element analysis data into a form which is suitable for graphical visualization.

PROFILING   (see POCKETTING).

PROTOCOL   The message framework within which data is transmitted down a communications line.

PUCK   An interactive input device which is placed on the surface of a graphics tablet or digitizer and whose position can be accurately sensed. Pucks often incorporate function buttons for various commands.

RASTER REFRESH DISPLAY   A display device which scans drawing data, on a line-by-line basis, from a bit-map memory onto the screen.

REFRESH   The process of continually re-drawing an image onto the screen of a raster display.

REGISTRATION   The degree of precision provided by an electro-mechanical output device such as a plotter.

REPAINT   Re-drawing of a display screen following editing.

RESOLUTION   A measure of the density of addressable points on a display screen. The higher the resolution the better the quality of the displayed image.

RESPONSE TIME   The total time taken to complete a requested operation at the CAD/CAM workstation.

ROBOTICS   Application of mechanical actuators and arms, under computer control, to perform manufacturing tasks.

SCULPTURED SURFACE   A surface generated by a modelling system to fit over a set of splines. This type of surface does not conform to rigid mathematical rules but is used extensively in body styling.

SELECTIVE ERASE   Un-drawing only specified parts of a displayed image, leaving the remainder intact.

SHADOW MASK   In a colour raster display, the shadow mask is a sheet perforated with fine holes through which the separate RED, GREEN and BLUE electron beams pass and focus on the phosphor coating of the screen. The mask keeps the beams apart.

SNAPPING   The action of automatically 'pulling' geometry to the nearest grid point or to other geometry on a drawing (see also GRID).

SOFTWARE   A term used to describe all programs used on a computer system.

SPLINE   Used in modelling, a curve which the system has fitted smoothly over a set of points. The user has control over the tightness of the fit and other characteristics.

STYLUS   Another term for an electronic pen.

SUBROUTINE   A set of program instructions which may be used at several different points in a program, or in several different programs.

SURFACE OF REVOLUTION   A surface model created by rotating (sweeping) a profile about an axis.

SWEEP   To produce a model by moving a profile through the third dimension by a given amount. This may be linearly or radially.

TRACKING   Following the position of an interaction input device and moving the screen cursor accordingly.

VIRTUAL MEMORY   A conceptual arrangement in which the user appears to have a vast working storage area, but in reality the operating system is paging (or reading) information from archival store at a rapid rate.

WINDOW A command which enlarges an area of a drawing for viewing and editing purposes.

WORK-STATION The CAD/CAM users environment, comprising a graphics terminal, interaction devices, a processor, and a communications link to the host system.

ZERO OFFSET A characteristic of an NC machine tool control permitting the zero point on an axis to be shifted readily over a specified range.

ZOOM (see WINDOW).

Index

Index